Hannes Griebel

Reaching High Altitudes on Mars
with an Inflatable Hypersonic Drag Balloon (Ballute)

VIEWEG+TEUBNER RESEARCH

Hannes Griebel

Reaching High Altitudes on Mars with an Inflatable Hypersonic Drag Balloon (Ballute)

VIEWEG+TEUBNER RESEARCH

Bibliographic information published by the Deutsche Nationalbibliothek
The Deutsche Nationalbibliothek lists this publication in the Deutsche Nationalbibliografie;
detailed bibliographic data are available in the Internet at http://dnb.d-nb.de.

This is a reprint of the dissertation with the title
'Reaching High Altitudes on Mars With An Inflatable Hypersonic Drag Balloon'
by Hannes S. Griebel

Academic title 'Doktor der Ingenieurswissenschaften' (Dr.-Ing.)
awarded on January 29, 2010.

Chairman:
Prof. Dr. Helmut Rapp

Board of Examiners:
Prof. Dr. Bernd Häusler
Prof. Dr. Christian Mundt
Prof. Dr. Hans-Joachim Gudladt

Universität der Bundeswehr München
(University of the Federal Armed Forces of Germany in Munich)
Department of Aerospace Engineering
Institute of Space Technology

1st Edition 2011

Editorial Office: Ute Wrasmann | Anita Wilke

Vieweg+Teubner Verlag is a brand of Springer Fachmedien.
Springer Fachmedien is part of Springer Science+Business Media.
www.viewegteubner.de

Cover design: KünkelLopka Medienentwicklung, Heidelberg
Printed on acid-free paper
Printed in Germany

ISBN 978-3-8348-1425-8

Artist's impression of an inflatable hypersonic drag balloon (referred to as "ballute"), flying through the outer atmosphere layers of planet Mars. The vehicle is shown around the point of maximum aerodynamic heating at 4.2 km/s and 90 km altitude.

Acknowledgements

The work described herein would not have been possible without the continuing assistance and support of the following institutes (in no particular order):

- The Mars Society Germany (Mars Society Deutschland e.V.)
- The Institute of Space Technology, UniBw München
- The Institute of Lightweight Structures, UniBw München
- The Institute of Thermodynamics, UniBw München
- The Institute of Material Sciences, UniBw München
- The Institute of Photogrammetry and Mapping, UniBw München
- The Institute of Statics and Dynamics of Aerospace Structures, University of Stuttgart
- The Institute of Space Systems, University of Stuttgart
- The Institute of Spaceflight, University of Applied Sciences, Bremen
- The Institute of Geology and Extraterrestrial Physics, University of Braunschweig
- The Finnish Meteorological Institute, Helsinki
- The Institute of Computer Science, Technical University of Iasi, Romania
- The Department of Electronics and Computers, University of Pitesti, Romania

The following companies and institutions have made major contributions to this development effort:
- The DLR MORABA group (reporting to the German Space Operations Centre GSOC at Oberpfaffenhofen) by launching flight test vehicles into space.
- The ESA Directorate of Human Spaceflight by providing parabolic flight time.
- IABG mbH at Ottobrunn, Germany, by supporting the project during assembly and testing phases.
- Lohmann Tapes at Neuwied, Germany, by developing and manufacturing a special high performance tape.
- Foster-Miller Inc. at Waltham, MA, USA, by manufacturing material samples.
- Die Firma J.Oed GmbH at Nurnberg, Germany, by providing their equipment to manufacture balloon segments.
- ARCOTEC through their cooperation in corona treatment processing and the provision of customized equipment.
- Olympiapark München GmbH by allowing a full scale inflation test at the Olympia Hall Munich.

In addition to these, a large number of students, private individuals and small companies are actively supporting the development effort by supplying their time, expertise, equipment, and hardware as well as personal and financial resources far beyond the ordinary.
Project ARCHIMEDES, for which this study has been made, is specifically tailored to include students of various majors in an effort to foster education in a real space project and to motivate young people to pursue careers in aerospace engineering and science.

I would like to take this opportunity to thank all contributors and team members for their outstanding work and the excellent time we had during this most exciting research programme!

Hannes Griebel

Table of Contents

List of Figures

List of Tables

List of Symbols

a_{albedo}	planet's albedo coefficient
a_d	velocity change resulting from drag (deceleration)
a_P	distance between the planet's centre and the point P on the terminator plane
$\alpha_{B,IR}$	ballute material's absorption coefficient in the infra red band
$\alpha_{B,VIS}$	ballute material's absorption coefficient
α_{PL}	xy-plane angle to trace of planet vector
α_S	xy-plane angle to trace of sun vector
A	area
A_H	Hull area of the ballute.
A_{sc}	drag effective face area of a spacecraft normal to the velocity vector
β	ballistic coefficient
β_{PL}	elevation angle of planet vector
β_S	elevation angle of sun vector
c_d	coefficient of drag
c_p	specific heat at constant pressure
c_v	specific heat at constant volume
$\vec{C}_{PL}$	vector to planet's geometric centre
C_{ij}	heat conductivity between two nodes
d_B	diameter of ballute
D_{fit}	fitting diameter
ε_B	ballute material's emission coefficient
ϵ_{TP}	terminator plane
F_d	drag force
F_g	gravitational force
F_{ip}	Instrument Pod pulling force
F_l	lift force
$F_{l,stat}$	static lift force
F_w	weight force
γ	ratio of specific heats
γ	flight path angle (FPA)
γ_{entry}	flight path angle at the atmospheric boundary (entry condition)
$\gamma_{SBTA,limit}$	Limit angle for atmospheric entries. Entries more shallow will not raise the sound barrier transition altitude any further.
$\gamma_{shallow,limit}$	Limit angle for atmospheric entries. Entries more shallow will lead to a skipping entry.
H	atmospheric density scale height
H_0	atmospheric density scale height of the lowest atmosphere layer (ground layer)
$H_{1,2...n}$	atmospheric density scale height of layers above the ground layer
H_P	atmospheric pressure scale height
Kn	Knudsen number
L	characteristic flow field dimension

λ	distance to heat source
λ	free molecular path
λ_{Pack}	packaging efficiency factor
λ_{pf}	ballute production factor
λ_s	safety factor
m_{BL}	mass of the ballute
m_{gas}	gas mass
m_H	mass of the ballute hull
m_{ip}	mass of the Instrument Pod
m_s	mass of a dimensionally equivalent sphere without seams, reinforcements and fittings
m_{sc}	mass of a spacecraft
n_{PL}	unit vector to planet
n_S	unit vector to sun
P	dissipated heat
p	pressure
P_P	point on the terminator plane
Φ	heat influx on a surface element
Q	heat
q_∞	free stream dynamic pressure
$\dot{q}_c$	stagnation point convective heating
q_{PL_i}	planetary infrared radiation influx
q_{PL_A}	planetary albedo radiation influx
$q_{SolarConst}$	local solar constant
R_{ij}	radiative heat exchange
R_g	gas constant specific to the inflation gas
R_{He}	gas constant specific to Helium
R_N	nose radius
R_{uni}	universal gas constant
ρ	density
ρ_∞	free stream atmospheric density
ϱ_A'	areal density of the ballute skin
ϱ_s	volumetric material density of the ballute skin material
σ	Skin stress.
T	temperature
T_g	gas temperature
t	time
t	layer thickness in a composite
τ_s	skin thickness
ϑ	trajectory parameter
U	internal energy
u	gas flow velocity
V	Volume
V_{BL}	Volume of the inflated ballute
V_{EnvMat}	Volume of the ballute envelope material
V_{PA}	Gas volume of protective atmosphere inside ballute container

V_{TA}	Gas volume of air trapped inside the ballute during packing
$\vec{v}_\infty$	free stream flow velocity
W	work
ξ	equivalent ballistic coefficient of displaced air parameter
ζ	weight parameter

List of Constants

$k_{M,V} = 18.9$

Gas mixture coefficient of the Martian and Venusian atmosphere for the stagnation point convective heating model by K.Sutton & R.Graves

$k_{air} = 18.1$

Gas mixture coefficient of the Earth's atmosphere for the stagnation point convective heating model by Sutton & Graves

$k_B = 1.38 \cdot 10^{-23} \dfrac{J}{K}$

Boltzmann constant.

$\sigma = 5.6704 \cdot 10^{-8} \dfrac{W}{m^2 K^4}$

Stefan-Boltzmann constant.

Acronyms and Abbreviations

ACRS	Air Cushion Restraint System (General Motors, 1973-1976)
AIR	Air Inflatable Retarder
AMSAT	Amateur Radio Satellite Organisation
AOCS	Attitude and Orbit Control System
ARB	Atmospheric Range Boundary
ARCHIMEDES	Areal reconnaissance Robot Carrying a High-Resolution Imager, a Magnetometer Experiment and Direct Environmental Sensors.
ARM	ARCHIMEDES Release Mechanism
ATR-FTIR	Attenuated Total Reflectance FT-IR
CamCon	Camera Module Control Computer
CCIR	Comité Consultatif International Pour La Radio
CCSDS	Consultative Committee for Space Data Systems
CEVCATS-N	Cell Vertex Central Averaging Time-Stepping Scheme-Nonequilibrium
CFD	Computational Fluid Dynamics
CVCM	Collected Volatile Condensible Material
DLR	Deutsches Zentrum für Luft- und Raumfahrt
DMA	Dynamical Mechanical Analysis
DOF	Degree of Freedom
EDL	Entry, Descent and Landing
EMU	Extra-Vehicular Mobility Unit
ESA	European Space Agency
FEM	Finite Element Method
FEP	Fluorinated Ethylene Propylene
FMI	Finnish Meteorological Institute
FPA	Flight Path Angle
FT-IR	Fourier-Transformed Infra-red Spectroscopy
GATG	Go Around To Ground
GATF	Go Around To Float
GRAMM	General Reference Atmosphere Model for Mars Missions
GSOC	German Space Operations Centre
HGA	High Gain Antenna
ICCS	Inflation Control Command Sequence
IDU	Inflatable Decelerator Unit
IGSS	Inflation Control and Gas Storage System
IGEP	Institut für Geophysik und Exterrestrische Physik
IR	Infra-red
IRS	Institut für Raumfahrtsysteme (Institute of Spaceflight Systems), TU Stuttgart
IRDT	Inflatable Re-entry and Descend Technology (Astrium / Lavochkin)
ISIC	In Space Inflation Concept
ITU	Internation Telecommunication Union
JPL	Jet Propulsion Laboratory
JPS	Joint Propulsion System
KEGSEC	Kopplung Euler Grenzschicht Equilibrium Version C

LEO	Low Earth Orbit
LGA	Low Gain Antenna
MCD	European Mars Climate Database
MIRIAM	Main Inflated Re-entry Into the Atmosphere Mission
MiriMag	Magnetometer Experiment for MIRIAM.
MMU	Manned Manoeuvring Unit
MORABA	Mobile Raketen Basis (Mobile Rocket Base)
NASA	National Aeronautics and Space Administration
NTO	Dinitrogen Tetroxide (Nitrogen Tetroxide or Nitrogen Peroxide), N_2O_4
PBO	Poly(p-Phenylene-2,6-Benzobisoxazole)
	or in short Polyphenylbenzobisoxazole
PEA	Pressure Equilibrium Altitude (PET Event Altitude)
PET	Pressure Equilibrium Transition
PFA	Perfluoroalkoxy (also know as Teflon®)
PodCon	Instrument Pod Control Computer
RAPHAELA	Radar and Photographic Altitude Experiment for
	Low-Power Applications
RCS	Reaction Control System
REGINA	Residual Gas Inflation Test for ARCHIMEDES
RML	recovered mass loss
RSS	Radio Science Simulator
SBTA	Sound Barrier Transition Altitude
ServCon	Service Module Control Computer
SSC	Swedish Space Corporation
TDS	Transportation and Deployment Subsystem
TML	total mass loss
TUM	Technical University of Munich
TÜV	Technischer Überwachungs Verein
UBW	University of the Federal Armed Forces of Germany in Munich
	(Universität der Bundeswehr München)
UDMH	Unsymmetrical Dimethylhydrazine, $C_2H_8N_2$
UHF	Ultra High Frequency
UV	Ultra Violet
VCM	Volatile Condensible Material
VHF	Very High Frequency

1 Introduction

The main purpose of this work was to find a way of placing a meaningful scientific payload within the upper atmosphere of Mars such that certain scientific measurements become possible. So the altitude range covered and the possible dwell time are relevant design drivers and will be explored in greater detail within this thesis.

Either concept of flying on Mars, however, be it heavier or lighter than air (read: heavier or lighter than carbon dioxide) is restricted to altitudes well below 10 km. Since any sustained flight at that altitude or higher in such a thin atmosphere is technically impractical, so is ascending to that altitude with a sounding rocket. Therefore a team of scientists from several research institutes around Europe formed a scientific committee. Together with engineers and scientists from the Mars Society Germany and the University of the Federal Armed Forces of Germany in Munich (UBW) the committee drafted a reference scientific mission scenario [1] to find ways to do exactly that: providing an instrument suite with access to a large altitude range and a meaningful dwell time (see 10.3.1 starting on page 208).

The most practical approach was found in an inflatable entry vehicle that can provide the largest and lightest possible configuration. Since no such attempts have yet been made, the work described herein focuses mostly on the mathematical theory and the technical peculiarities of such a system, but also treats mission design and testing aspects.

1.1 Scientific Background

Until the inception of interplanetary space flight, the science of atmosphere physics and climatology was forced to restrict its research more or less to only one specimen: the atmosphere of planet Earth. More because attempts at investigating planetary atmospheres using spectroscopy with ground based observations have been made and less because some regions of the Earth's atmosphere remained beyond reach before the advent of rockets.

Because Mars is the planet which resembles our own most closely, it is of special interest to planetary science [2]. It is the only body of which we can say, with sufficient confidence, that it has once had a climate similar to Earth's and may even have harboured primitive forms of life. Today, however, Mars is a barren, cold and dry desert planet with only a ghost of an atmosphere left. The question is thus why Mars underwent such a vastly different development, when conditions at the outset were so similar.

This is why until today the Martian atmosphere has been a subject of intense study, most notably by radio sounding from satellites such as presented by references [3] and [4]. This method has contributed most of the data and our present understanding of the Martian atmosphere.

In addition to atmospheric research, high altitudes (or the traversing of a great altitude range during a meaningful period of time) has distinct benefits for other research areas as well, as the field of geomagnetism so tellingly demonstrates [1]. Placing a suite of scientific instruments on a trajectory travelling through an atmosphere across extended geographic and altitude ranges is therefore an important method of research.

1.2   Engineering Background

Ascending from the surface of Mars with a balloon is difficult at best, because relatively high wind speeds thwart the deployment of large gossamer balloon hulls or aeroplane wings without considerable mechanical effort [5]. Besides, the altitudes which such a system can ascend to are not very high. A high altitude sounding rocket commonly used on Earth in contrast might reach all regions of the Martian atmosphere, however, is quite unavailable. The large mass and complexity of such a system renders it unsuitable for a launch from the surface of a remote planet.

Instead of ascending from the surface, however, a high altitude mission might also descent from space, provided it can be built to decelerate at a sufficiently high altitude to make scientific gains desirable. No conventional hypersonic aeroshell and parachute system will slow down sufficiently to allow deployment far enough above the surface, because its mass to drag ratio (expressed in the so-called "ballistic coefficient", see chapter 3.4) is much too high.

In summary, one can state that the effect of a lower ballistic coefficient raises the entire deceleration profile to a regime where the atmosphere is less dense (a higher altitude), prolongs descent time and lowers the aerothermodynamic heating load that the vehicle will have to absorb. This translates into the following capabilities:

- Increasing payload dwell time in the atmosphere during descent.
- Increasing the accessible altitude range for direct measurements.
- Landing at a site with such a high altitude that heavier and speedier probes cannot land on them because they cannot decelerate in time for a safe landing (such as Tharsis on Mars).
- Allowing a spacecraft to obtain an almost fuel-free delta v through aerobreaking.

In order to achieve the highest possible altitude profile, the ballistic coefficient has to be as low as technically practical. This is achieved by using a large and light inflatable sphere, which in itself offers some benefits as well. Since it can be densely packed and deployed with a small number of moving parts, it offers an attractive alternative to rigid structures. So the same technology might also be considered for the following applications:

- Providing a large aerodynamic heat shield for a heavy payload when a rigid heat shield is too big for the biggest rocket payload fairing.
- Providing an inflatable solar sail structure

While the main focus of this work is a specific task on Mars, chapters 10.4.1 and 10.4.2 give two more detailed examples for possible other applications.

2 Related Technologies and State of the Art

2.1 About Planetary Aeroplanes and Balloons

Despite the large number of planetary probes that have successfully completed orbital and landing missions on Mars [6], we find ourselves left with a gap in the accessible measurement range: That of the intermediate scale. That of long range mobility of planetary dimensions combined with in-situ atmosphere measurements and birds-eye perspective imaging. A very good overview of past concepts, actual attempts and future plans for planetary areal reconnaissance missions can be found in reference [7].

Two methods exist to suspend a vehicle within the atmosphere of a planet: The dynamic lift method (for heavier-than-air designs) and the static lift method (lighter-than-air). Fixed or rotor wing craft generate lift through air that creates a pressure difference when it flows over an air foil and are comparatively small and heavy. Alternatively, a parachute or parafoil may be used to slow the descend speed of a heavier-than-air object by sheer drag. Balloons and airships in contrast are big and light enough to displace atmosphere gas of greater mass than their own vehicle mass. This mass difference can be used by a payload. Because the vehicle creates buoyancy and does not need air flowing around it to generate lift, it is also referred to as "static" lift.

Lighter-than-air vehicles can again be divided into two main categories: Zero-pressure designs and overpressure designs. The former are in contact with the surrounding atmosphere (usually through an opening at the bottom) and therefore maintain pressure equilibrium with the surroundings. The lower density of the inflation gas (often enough the same as the ambient gas in which the vehicle floats) is created through heat. A typical example would be a hot-air balloon. An over-pressure design on the other hand has an inflation gas which is lighter than the atmosphere in which it floats. It maintains the shape of the envelope because the inflation gas has a higher pressure than the ambient atmosphere. Helium-filled weather balloons are a typical example. A special type of an overpressure design is a dirigible (also referred to as airship), which is oblong and streamlined in shape and can thus be more easily moved around in a controlled manner.

Planetary probe studies based on either concept are also mentioned by reference [7]. Only a very small number of airborne planetary missions have been flown to date, however, but those who did were rewarded with outstanding success. On Venus, which has a very dense atmosphere, numerous Russian landers (the "Venera" series) and American probes ("Pioneer Venus") have parachuted through the thick atmosphere and transmitted sensor readings over large altitude ranges. Venus has even seen the deployment of two Franco-Russian Helium-balloons (VEGA-1 and VEGA-2), which have travelled great distances [7].

On Saturn's moon Titan, only the European probe Huygens has descended to the surface, if only beneath a parachute. But this mission has already revolutionized our understanding of this complex Saturnian world.

No dedicated atmosphere probe has yet been sent to Mars. That doesn't mean that no suggestions were ever made to reach Mars with aerostats and aeroplanes. Ever since the first probes reached Mars have researchers of many disciplines expressed their interest in high altitude research for Mars and drafted concepts. A French hot air balloon was a suggested payload to the Russian Mars 94 probe [7] and was the one project of such sort that got closest to actual deployment on the planet (it was cancelled along with the Mars 94 mission very late in the development phase). A vast number of aeroplanes of different shapes and sizes, such as the one presented in [8], were suggested as payloads to Mars missions, along with an even larger number of dirigibles and balloons. One of the most notable and thorough efforts of an airborne reconnaissance probe development efforts for Mars was the MABVAB programme [9]. Restricted to below 7 km, however, all suggested systems were more geared towards covering a large distance for areal survey than for high altitude research.

Despite the many suggestions and the high scientific interest, the VEGA Venus balloons remain the only planetary balloon mission ever realized. Part of the reason is that flying in Venus' thick atmosphere is comparatively easy.

Mars, in contrast, poses a more complex challenge. At a mere 7 hPa average surface pressure, it is certainly thick enough to necessitate a hypersonic entry vehicle and a supersonic parachute system, but unfortunately not thick enough to use these for gently descending to the surface. Ordinary recovery systems are unable to slow a vehicle to reasonable landing speeds, so airbags or retrorockets have to be used for a soft touch down. In short, any extended dwell time in the Martian atmosphere becomes a technical challenge that hasn't yet been met. It takes just six minutes for an ordinary lander to reach the surface and it is not until the last minute that most of the crucial landing events have to take place. This fast pace leaves very little time for any atmospheric instrument to grasp readings during the last couple of kilometres above the surface. It is small wonder therefore that high altitudes on Mars have remained inaccessible to instrument carriers so far. Moreover polymers able to sustain expected thermal and mechanical loads acting on the envelope of a Mars balloon have only become available in the last two decades.

On Titan, sustained flight or floatation would be possible with moderate effort, however, Titan's remote location in the Saturnian system makes it very hard to reach.

No heavier-than-air dynamic lift creating vehicles have ever been flown in an atmosphere other than Earth's.

2.2   About Automatic Inflatables

Automatically inflatable structural elements are commonly used today in a large number of applications. Some of the most prominent are automobile airbag systems, emergency life vests and inflatable rafts on modern ocean liners and aeroplanes. Most of these systems rely on chemical gas generator cartridges, however, compressed gas systems and hybrid systems have also been used in a number of applications.

2.2.1   Gas Generator Systems

Gas generator systems can come as cartridges that are ignited and produce gas in an instant, or as sublimation systems which generate gas over a period of several tens to hundreds of hours. Cartridges are used in most automated inflation applications today. For example, automobile inflatable restraint systems (airbags), emergency flotation devices and evacuation slides use this method exclusively. Some space inflatables, such as the Echo balloons [10], relied on sublimation systems [11][12]. But other space inflation systems, most notably radar tracking and calibration spheres, the inflatable re-entry technology demonstrator IRDT [13] [14] and landing cushions used on Mars all used cartridges as well.

The main advantage of the gas generator cartridges is that the gas is chemically bound to another compound, making it easily storable under varying ambient conditions for decades. Since no pressure vessel is needed, the danger of loosing gas through system leakage does not exist. For the same reason, tank rupture through fatigue is impossible and containment weight is very low. Releasing the inflation gas is achieved quite simply by triggering the chemical gas generation process, electrically or otherwise, and inflation gas is available instantly and reliably.

Sublimation systems on the other hand are subjected to ambient conditions that lead to the decomposition of the compound. The development programme of the ECHO balloons showed, with its many failures, that this process is difficult to control and can easily lead to hull rupture, if the process cannot be regulated or at least be halted. An advantage, however, is that sublimation systems decompose over many days and can therefore constantly replenish small amounts of inflation gas which might vent through envelope damages.

A major drawback is also the limitation of available inflation gases to those which can be bound to other chemicals such that the resulting substance is stable under all expected mission conditions for the entire expected mission time. Another is the resulting impurity of the inflation gas and the relatively imprecise prediction of the amount of available inflation gas, since the decomposition of the gas generation charge is never one hundred percent complete. For the same reason, by-products from the reaction are left inside the inflatable, and heat generated in the process is transferred to the inflatable's skin. In case of a triggered cartridge, inflation gas is generated almost instantly and uncontrolled, so the deployment process can only be controlled by the packaging pattern of the inflatable. Sublimation systems in turn may take many tens of hours to produce the full amount of gas and are difficult to stop once initiated.

The application of an atmospheric sounding ballute with the requirement of a long and defined descent time makes the application of a heavy inflation gas mixture with a poorly known amount of bi-products unfavourable. The fact that the ballute envelope should be as light as possible and therefore delicate and fragile in unpressurised state (especially in case of the Mars mission) makes the almost uncontrollable inflation process of a gas generator system unattractive. Furthermore, rapid inflation is neither necessary nor desirable.

2.2.2 Compressed Gas Systems

The first commercially available automobile airbags, though ultimately not economically successful, were driver and passenger airbags offered by General Motors as optional equipment on the '74 to '76 models of full-size Cadillacs, Buicks and Oldsmobiles. These systems were based on compressed air held in pressure vessels [15]. Since this system offers some interesting solutions, it is explored in more detail in chapter 2.2.4.

Another noteworthy example is the inflatable antenna experiment (IAE) on Space Shuttle Endevour's mission STS77. The IAE was developed by L'Garde, Inc. of Tustin, California, under contract to NASA and tested the deployment of a light and gossamer, yet geometrically precise, space inflatable structure [16].

Compressed gas systems offer much better control over the inflation gas, at the prize of added complexity, weight and volume. Virtually any gas can be stored in compressed state, albeit not necessarily for any length of time. They can also be used many times by refilling the tanks – an important test criterion – and the amount and composition of the inflation gas can be precisely controlled. Therefore, compressed gas systems are preferable for systems that require a precise amount and composition of inflation gas along with good controllability of the inflation process. Another handy advantage is that compressed gas is available at any desired quantity during the mission. Compressed gas, especially Helium, is needed anyway on most spacecraft, be it to pressurize fuel tanks, cold gas thrusters or flood volumes which require an atmosphere to function.

2.2.3   Cryogenic Gas Storage Systems

For the sake of this analysis, cryogenic gas storage systems have also been briefly considered. Storing Helium cryogenically can significantly reduce its volume and no pressure vessel is required. However, long-term boil-off rates make this method impractical for the intended application. In addition to this, a rather complex active thermal control system is required, which adds a number of possible failure modes and a permanent power consumption to the system.

To the knowledge of the author, no inflation system ever existed that stored inflation gas this way.

2.2.4   General Motors' Air Cushion Restraint System from 1973 to 1976

General Motors' Air Cushion Restraint System (ACRS) is interesting because it is the only mass produced automobile airbag system relying on compressed gas for inflation and because it has a two stage inflation system that pressurized the air cushion depending on the severity of the impact. It is also unique in that it relied on both compressed air and gas generator cartridges to control the inflation process and inflation gas mass.

Fig. 2.1: Inflated driver side ACRS air bag [15]. (GM)

Airbags were subject of extensive development efforts for at least a decade and a half until General Motors perfected the first such system to be installed on mass produced vehicles. The system was dubbed Air Cushion Restraint System (ACRS). A test run was made on 1973 Chevrolet Caprice and Impala models intended for government use. Starting with the 1974 model year, driver and passenger airbags of that design were available as optional equipment on General Motors' full size models [17].

The ACRS consisted of two modules: a simple single stage system for the driver installed in the steering wheel hub (Fig. 2.1) and the pressure bottle fed two stage system for both front passengers installed above the right passenger's knees (see Fig. 2.2). The inflation gas for the driver came from a gas generator cartridge. The bulk of the inflation gas for the passenger ACRS was stored in a pressure vessel mounted behind the air cushion compartment, complemented by two individually controlled gas generators. The inflation was triggered by an electronic control module monitoring a complement of accelerometers mounted behind the front bumper and inside the module itself. If direction and magnitude of a collision event exceeded predefined limits, the control module would trigger pressurization of the air cushions [15]. Depending on the severity of the impact, either one or both gas generator cartridges were triggered to boost the pressure bottle's initial charge.

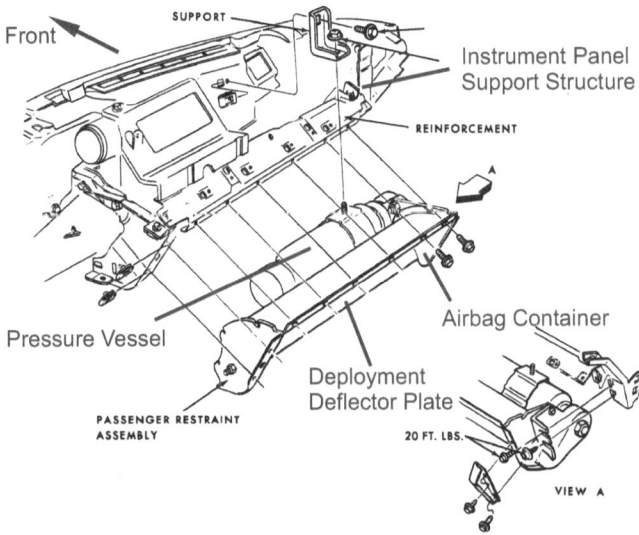

Fig. 2.2: Sketch of the passenger side ACRS assembly [15] (Cadillac).

The deployment kinematics of the air cushion was controlled through the packaging pattern and the shape of the rear deflector plate (see Fig. 2.2), a technique still common for today's gas generator based air bags.

The packaging efficiency (achievable volume reduction through packing) was low, but size was a lesser constraint than reliable deployment kinematics. A general problem of compressed gas inflation systems is the bulky pressure vessel. Together with the low packaging efficiency of the large twin-passenger air cushion, it ate away much of the front passenger legroom although it already replaced the entire glove box. The location of the packed passenger-side air cushion module can be seen in Fig. 2.2. Today huge advances in packaging techniques, computer simulations of airbag deployment kinematics and mass produced gas generator cartridges have made airbag modules cheap, light and small without sacrificing reliability (unfortunately, the high level of specialization of these tools render them unsuitable for predicting the ballute deployment of our design in space).

Another typical feature of automobile airbags can be seen in Fig. 2.1, namely that of a two segment design. A more or less conical "under-body" is attached to a more or less spherical canopy through a single equatorial seam. This is practical for airbags, since their skin material is rather elastic. However, wrinkles around the seam show that the inflated cushion departs from its natural shape significantly. These are areas of considerable stress and strain under pressure. While this is acceptable for airbags with no real weight limit, a ballute constrained by weight and an aerodynamically smooth surface must be made with enough segments to minimize excessive strain and wrinkling.

In terms of tightness, it is further noteworthy that in 1993, two '73 Chevrolet Impala models manufactured in late 1972 and equipped with ACRS were crash tested by the Insurance Institute for Highway Safety [18]. Both vehicles were in a much deteriorated state from sitting unused and unprotected for a number of years. In both vehicles, the ACRS deployed correctly, showing that the pressure vessel could hold the charge for twenty years without significant leakage, and that the impact detection and inflation control system could survive the same period of time without maintenance.

2.3   About Space Inflatables

A number of inflatable structures have been flown in space, ranging from inflatable structural elements such as antennas and trusses to entire inflatable space habitats to test their suitability as space station modules [19].

The most interesting series of such space craft are the Echo I and II balloons (see Fig. 2.3) and their development programme during the 1960s [20]. The Echo satellites were NASA's first attempt at building communications satellites. They were intended to be fully passive by providing a large metallized balloon to bounce communication signals off its surface. The spherical shape was chosen primarily to eliminate the need for active pointing. Echo 1A

Fig. 2.3: The 135-ft Echo balloon satellite during a static inflation test [11]. (NASA)

(commonly referred to as Echo 1) was successfully tested with intercontinental telephone, radio and television signals. This vividly proved the feasibility and worth of communication satellites, but already provided ultra high altitude atmospheric data in the process. Despite their orbital altitude between 1000 and 2000 kilometres, the large and light Echo balloons were extremely sensitive to residual atmospheric drag and solar wind interaction. Echo 1A re-entered on May 24, 1968 followed by Echo 2 on June 7, 1969 after only 4 years in orbit. This very well demonstrates the atmospheric drag effect of large and light objects. Since neither balloon was designed to withstand the atmospheric entry loads, both were destroyed during hypersonic flight.

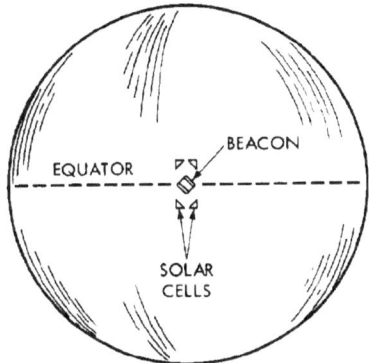

Fig. 2.4: Configuration of the Echo balloons [11]. (NASA)

Echo 1 was a 30.5 m diameter balloon made of aluminized mylar polyester thin film. Two 107.9-MHz beacon transmitters with solar cells were integrated into the balloon skin for telemetry, located 180° apart on the equator (see Fig. 2.4). Echo 2 had a 41.1 m diameter mylar envelope that used an improved sublimation inflation system [11]. Instrumentation included temperature and pressure sensors to monitor the balloon's skin temperature and mechanical loads. Like Echo 1A, Echo 2 carried a set of two beacon transmitters powered by solar cells and NiCd-rechargeables.

Because of their high relevance for the prevailing design, further reference on particular details is made where appropriate.

2.4 About Hypersonic Inflatable Drag Devices

Tailoring the ballistic coefficient of a vehicle to suit a specific mission adds a degree of freedom to the mission design process. For some applications, space vehicles with very low ballistic coefficients offer attractive mission improvements over conventional spacecraft, or sometimes even make certain mission profiles possible that are otherwise not. The concept of using a balloon to alter the aerodynamic properties of a space vehicle for atmosphere manoeuvres was most prominently described in 1982 by Arthur C. Clarke in his novel "2010: Odyssey two", where a spaceship enters the atmosphere of planet Jupiter with a set of ballutes to bring the spaceship onto an intercept trajectory with another spaceship adrift in the Jovian system. In the story, the ballutes are discarded after the aerobraking phase. A most dramatic rendition of this manoeuvre can be watched in the 1984 motion picture "2010: The Year We

Made Contact". The original idea, however, is much older and was first investigated by the Goodyear Corporation in 1959 (see chapter 2.4.1).

Today, the term "ballute" is used for any inflatable device intended to raise an object's aerodynamic drag, no matter the shape or intended purpose. But that wasn't always the case. The term "ballute" was first coined by the Goodyear Aerospace Corporation. The device, discussed in greater detail in chapter 2.4.1, was an almost spherical object that really had a notable resemblance to a balloon, intended to replace a conventional supersonic disc-gap-band style parachute in regions where either the speed of the vehicle or its altitude were considered too high for the reliable deployment and effectiveness of the latter [21][22].

One such application was the recovery system on the high-altitude ejection seats of the Gemini spacecraft, which used a ballute of Goodyear's design to recover astronauts forced to eject at great altitudes.

2.4.1   The Goodyear Ballute

While investigating technologies for high-speed decelerator systems in the 1960s, the Goodyear Corporation designed a balloon-shaped parachute which they named"Ballute" by synthesizing the two terms into a new one.

In contrast to the scope of the research effort described herein, the Goodyear ballute was designed for high altitude and high speed aerodynamic deployment, rather than inflating the ballute in space to use it for atmospheric entry. It was designed as a recovery system for Mach numbers exceeding those of conventional parachutes. Since directional stability was a requirement, the optimal shape was found to be a conic forward section with a spherical aft dome. The result looks similar to a punching bag turned upside down (see Fig. 2.5). The conic forward section adds directional stability at the expense of skin stress, weight and increased heat load as compared to a simple sphere.

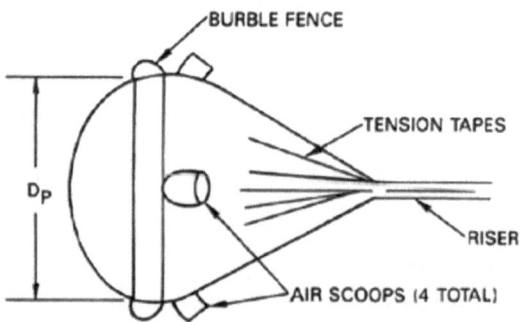

Fig. 2.5: Sketch of the Goodyear Ballute configuration [22]. (Goodyear)

A burble fence, a toroidal bulge around the ballute, further adds directional stability by ensuring a uniformly separation of the boundary layer, minimizing tumbling and roll. On a perfect sphere in contrast, asymmetric vortex shedding causes the ballute to roll and tumble.

The ballute was initially designed to be inflated through gas generator cartridges or compressed air bottles. Later in the development, however, four ram-air scoops were added instead, providing an adequate inflation pressure by directing oncoming air into the ballute. That way, a separate inflation system could be spared [21][22].

While the Goodyear Ballute serves a completely different purpose than the deceleration of vehicles coming in from space at astronautically relevant speeds, this ground breaking work laid many foundations upon which modern ballute technology draws. It also shows that the idea of the ballute itself is not new.

2.4.2   The Gemini Ejection Seat Recovery System

Apart from the test items built by Goodyear, ballutes were most notably used as a recovery system on the high-altitude ejection seats of the Gemini spacecraft [23].

Fig. 2.6: Recovery test with the Gemini ejection seat ballute [23](NASA).

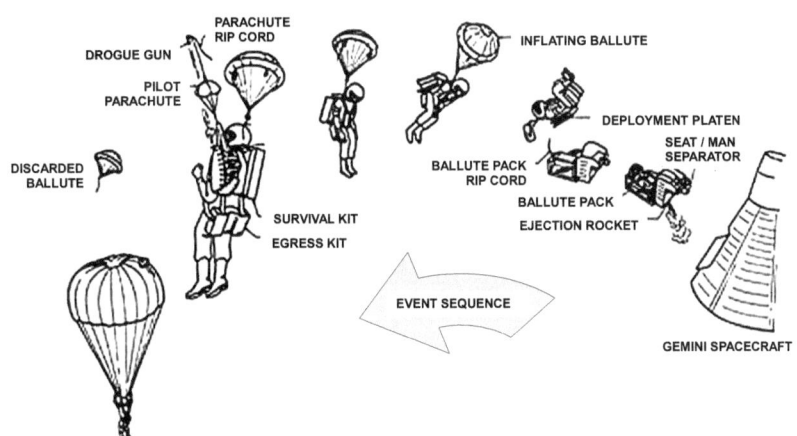

Fig. 2.7: Gemini recovery ballute deployment sequence (adapted from [23]). (NASA)

The ballute was intended to stabilise the astronaut after ejection from the spacecraft (Fig. 2.6). It was deployed from the ejection seat back pocket and inflated through ram air inlets. At lower speeds and altitudes, the ballute was foreseen to be jettisoned and a conventional recovery parachute deployed by a drogue gun and pilot chute was used for landing (Fig. 2.7). Initial drop tests with dummies showed several problems with the ram air inflation at subsonic speeds, whereas at supersonic speeds the ballute would often fail. Additional roll and pitch stability issues led to a redesign of the system, which resulted in a series of very successful dummy and test pilot drops from between 12000 and 35000 feet (Fig. 2.6). Initially. the tested ballutes had a diameter of 42 inches (ca. 1m), which was later increased to 48 inches (ca. 1.2m) as part of a redesign.

2.4.3   IRDT

The inflatable re-entry and descent technology demonstrator (IRDT) flown into orbit and back on February 9th, 2000 was the first atmospheric re-entry from space using an inflatable decelerator [13]. A consortium of Lavochkin and Astrium Bremen cooperated on the project after Lavochkin's earlier attempt at flying a similar entry vehicle on Mars-96, which perished on route to Mars.

IRDT was tested a total of two times and, although neither test performed as intended, a lot of experience could be gained. IRDT lowered the ballistic coefficient (see chapter 3.4) by one order of magnitude to somewhere on the order of 10^0 as compared to classic entry vehicles (on the order of 10^1).

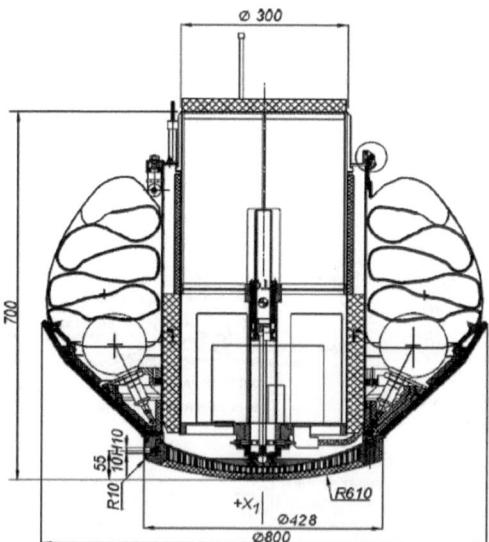

Fig. 2.8: Sketch showing cut through the stowed IRDT-2 test vehicle. [13] (Astrium)

The IRDT test vehicle is shown stowed in Fig. 2.8 and deployed in Fig. 2.9. It had an equipment container and the inflatable decelerator referred to as IDU (Inflatable Deceleration Unit). The IDU was made of inflatable radial toruses that were interconnected by fabric sheets. The IDU could be inflated in two stages: the first stage (main stage) acted as an aerodynamic heat shield, while the second stage increased dynamic drag after peak heating, replacing a classic parachute system.

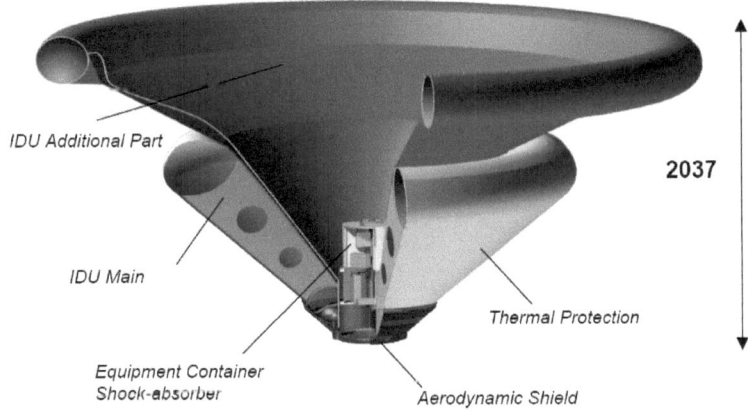

IDU Additional Part

2037

IDU Main

Thermal Protection

Equipment Container
Shock-absorber

Aerodynamic Shield

Fig. 2.9: Conceptual design of IRDT-2. [13] *(Astrium)*

2.5 State of the Art

Today, studies of ballute applications exist aplenty. Studied ballute shapes do not only include spherical objects akin to normal balloons, but also toroids, cones and lentils [24][25][26]. Studied applications are mostly inflatable drag bodies intended to replace rocket engines for orbital manoeuvring, promising large savings in vehicle weight.

Practical ballute applications today exist in only a very small number and all of them are more or less iterations of the Goodyear design. None have as of now been used to fly a dedicated space mission and no sealed, super pressurized drag balloon has yet been flown. A typical ballute application is the ram air inflated retarder on the high drag version of the MK83 bomb, also referred to as MK83-AIR (see Fig. 2.10).

More recent studies also revealed that the rather cumbersome toroidal burble fence bulge, which is costly to manufacture and quite problematic to attach, may be replaced by much simpler ribbons or taped slats, which shed vortexes just as effectively. Such tapes are easily added and since long state of the art in other applications requiring vortex shedding, such as modern sail planes.

Automatic deployment and inflation systems, however, exist in large numbers. Most are simple single stage gas generator based systems, but compressed gas systems exist and even combined gas generator / compressed gas systems with multi-stage inflation control systems have been built and used successfully.

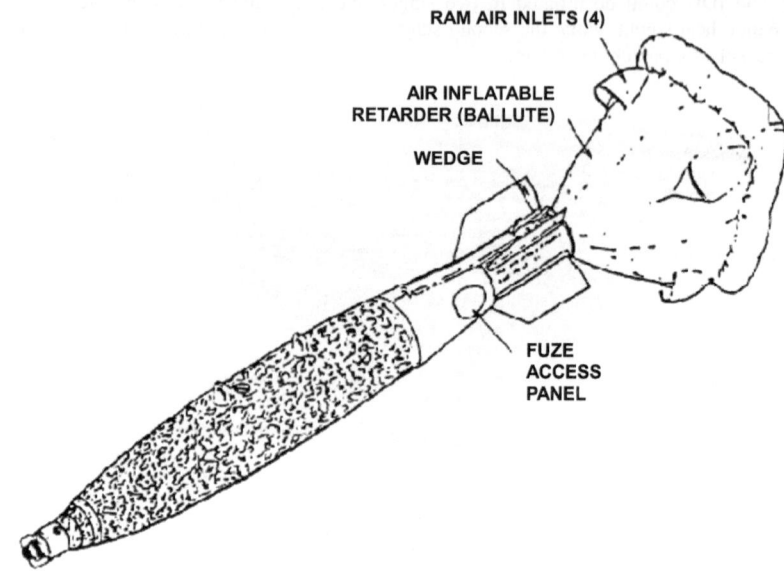

Fig. 2.10: Ballute configuration on the Mk 83 bomb (adapted from "http://www.ordnance.org/gpb.htm", 2009).

3 Basic Considerations on Probes with Low Ballistic Coefficients

To understand the technical concept behind an atmospheric sounding ballute for Mars, it is necessary to afford some attention to the peculiarities of such a configuration when it comes to atmospheric entry.
Two very good textbooks on the general subject of atmospheric entry are references [27] and [28].

3.1 The Spacecraft Elements

3.1.1 The Ballute Spacecraft

In order to facilitate scientific measurements over a larger altitude range and a considerably longer measurement time, a ballute offers a practical solution. The design suggested herein uses an over-pressure type spherical balloon as an inflatable hypersonic drag device. Through

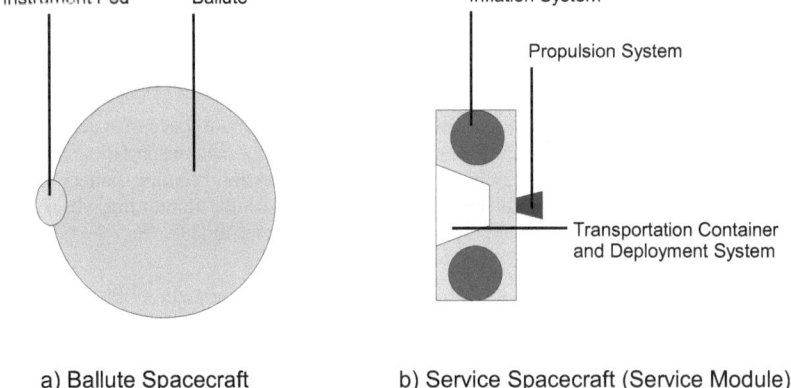

a) Ballute Spacecraft   b) Service Spacecraft (Service Module)

Fig. 3.1: The generic ballute spacecraft and service spacecraft

its over-pressure design, it combines the aerodynamic drag of a parachute with the static lift of a Helium balloon. The configuration remains unchanged throughout the entire critical mission phase of entry, descent an landing (EDL), thus making complex actions unnecessary and reducing the risk of mechanism failure to a minimum.

We take the liberty to define as a ballute spacecraft every space vehicle intended to enter a planetary atmosphere with a ballute, whereby the term ballute refers to any inflatable object designed to significantly influence the vehicles aerodynamic properties. We then divide it into two main parts: the instrument pod (or payload pod) and the ballute itself (see Fig. 3.1 and also Fig. 4.1).

The instrument pod is the part which carries the payload and other subsystems of the spacecraft. In case of a sounding probe, this is where the scientific and other flight instruments are located. But the pod can also act as a fairing to deliver a landing system or to recover objects from space and may then be referred to as "payload pod". Since the main purpose of this study is to investigate the intended use as an atmosphere probe for Mars, the term "instrument pod" is used throughout the entire text.

The pod is therefore a dense, rigid and hard object, with a structure made primarily of metals or rigid composites.

The ballute on the other hand is a large and light object, gossamer as compared to the pod, with a thin and most likely flexible skin made from some fabric or film or a laminate thereof.

Note that these definitions are used regardless of spacecraft configuration and shape.

We can define the mass budget of a generic ballute spacecraft as the sum of the instrument pod mass, the ballute hull mass and the inflation gas mass:

$$m_{sc} = m_{ip} + m_H + m_{gas} \tag{1}$$

3.1.2   The Service Spacecraft and Mission Infrastructure

To deploy a ballute spacecraft it has to be transported, along with its inflation gas, to its initial position in space. After successful deployment, the ballute needs a communication data link. We therefore define as Service Spacecraft (or Service Module, whichever seems more appropriate) all subsystems required to reliably deploy and inflate the ballute spacecraft according to mission requirements.

The service module contains all systems directly related to the ballute spaceraft's deployment. This includes the inflation control and gas storage system, the ballute transportation container and auxiliary subsystems. Such auxiliary systems could possibly contain battery charge electronics, equipment to provide for cruise communication, health monitoring, observation equipment targeted at the ballute, equipment to remove the service module from the ballute (if so desired), a propulsion system and other functions.

Under the terms "mission support infrastructure" or "flight infrastructure" all systems are grouped that help the ballute spacecraft and its service module to perform as intended. This includes a transportation vehicle and the provision of a reliable data or payload retrieval system.

3.1.3   Basic Simplification

For simple spacecraft configurations and parametric studies considered in this analysis, the following simplifications may be assumed, which hold reasonably well with low ballistic coefficients.

As a basic simplification, the volume of the instrument pod is always much smaller than that of the ballute and anyway not protruding significantly from the ballute shape. This also means that the shape of the spacecraft is mainly that of the basic shape of the ballute. Or in other words, protrusions of the instrument pod are small in size and aerodynamic effect as compared to the ballute.

This translates into the important simplification that the spacecraft volume is approximately that of the ballute and the spacecraft's overall coefficient of drag is also roughly that of the ballute, so that:

$$V_{spacecraft} \approx V_{BL} \tag{2}$$

and

$$c_{d,spacecraft} \approx c_{d,ballute} \tag{3}$$

This is of course not true if either a more detailed analysis is desired, or the configuration of the ballute spacecraft obviously violates that assumption too much. But such cases will be treated separately, where appropriate. For a spherical ballute intended to function as an atmospheric sounding probe on Mars, this assumption is very good.

3.2 Basic Equations of Motion

The basic equations of motion, upon which all analyses herein are based, can be found in [27] and [28] and represent the sum of all forces acting on the spacecraft. We choose a vehicle-fixed coordinate system oriented such that it points with its x-axis in flight direction and with

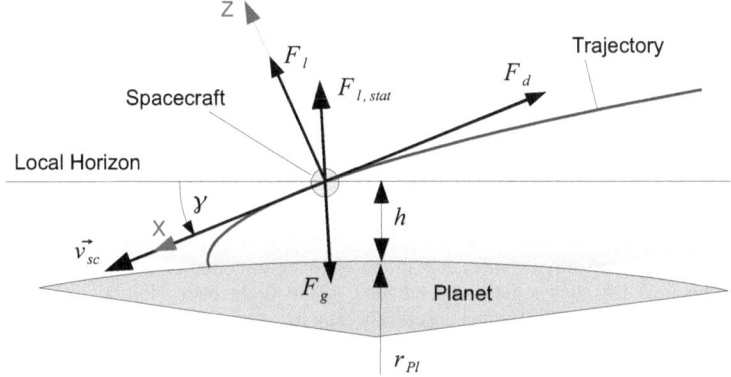

Fig. 3.2: Basic planar coordinate system and forces.

its z-axis perpendicular to it, away from the surface. The two dimensional (planar) case is depicted in Fig. 3.2. In this case the sum of all forces amounts to

$$m_{sc} \cdot \frac{dv_{sc}}{dt} = -F_d + F_w \cdot \sin \gamma \quad \text{in x direction (along flight direction)} \tag{4}$$

and

$$m_{sc} \cdot \frac{v_{sc}^2}{r_{sc}} = F_l - F_w \cdot \cos \gamma \quad \text{in z-direction (perpendicular to flight direction).} \tag{5}$$

Here subscript l denotes lift, d drag and w the vehicle weight. Subscript sc refers to the spacecraft, where r_{sc} is the distance of the spacecraft from the planet's centre and therefore

$$r_{sc} = r_{PL} + h \tag{6}$$

Because we are considering a ballute of sufficient dimensions to create notable static lift, it must be included in the calculation of the weight force. The weight force is thus the vehicle's downward force due to gravity diminished by the static lift force of the ballute body:

$$F_w = F_g - F_{l,stat} \tag{7}$$

The static lift force is calculated using the general equation for static lift [29]:

$$F_{l,stat} = \rho_\infty \cdot g \cdot V_{Bal} \tag{8}$$

where ρ_∞ is the atmospheric ambient density, g the local gravitational acceleration and V_{Bal} the ballute volume.

Dynamic lift and drag forces are modelled using standard equations [27] where the drag force is

$$F_d = \frac{1}{2}\rho_\infty v_{flow}^2 c_d A_{sc} \tag{9}$$

and the lift force is

$$F_l = \frac{1}{2}\rho_\infty v_{flow}^2 c_l A_{sc} \tag{10}$$

The coefficients c_l and c_d are the coefficients of lift and drag, respectively.

Note that the static lift force always acts in the opposite direction of the vehicle weight, but that the dynamic lift always acts perpendicular to the flight path. Note also that the flow velocity v_{flow} is not identical to the spacecraft velocity v_{sc}, unless we opt to neglect wind and planetary rotation. For all approximate analytical methods, this may safely be assumed, however.

For the full six-degree-of-freedom model, the equations of motion have to be expanded to account for cross-track deviation due to wind, planetary rotation and the spacecraft rotation. If the vehicle rolls to one side, the lift vector gets a y-component and thus causes a cross-track deviation. An extensive and very good treatment of full six-degree-of-freedom models can be found in reference [30].

3.3 About Atmosphere Models and Their Suitability

Before we begin any analysis in greater detail, it is practical to review available atmosphere models and their suitability for high altitude entry studies. A host of models exist for Earth and planetary atmospheres, most of which are given in large tables or even larger databases. A very good overview of available models is given by reference [31]. For the study of Mars entries herein, the European Mars Climate Database (MCD) [32] was used, because it is the most complete database to date. It therefore allows sensitivity analyses and the impact of seasonal and daytime atmospheric variations on the mission, thereby also allowing a study of expected trajectory dispersion Monte-Carlo-analyses and the derivation of navigational accuracy requirements.

The implementation of such an elaborate atmosphere model, however, requires a lot of work. For a quick look analysis, such as a back-of-the-envelope estimate or application in a concurrent design facility, it might be more practical to extract a typical atmosphere profile and use it in tabular form. When researching analytical solutions (such as in chapter 7.2.1), it might even be favourable to have an analytical atmosphere model at hand. A good model is a simple exponential relation between density and altitude. To understand its effect on mission analysis and its usability limits, we will briefly explore the underlying theory.

Consider for this purpose that the density and pressure in an atmosphere layer depend on the weight of the atmosphere above it and that the relation between pressure and density is that of an ideal gas. This results in the following relations for the density and pressure in terms of the height h from the bottom [33]:

$$\rho = \rho_0 \cdot e^{\frac{-h}{H_0}}, \quad p = p_0 \cdot e^{\frac{-h}{H_p}} \tag{11}$$

where ρ and p are local ambient density and pressure and ρ_0 and p_0 the atmospheric density and pressure at the bottom of the layer where $h=0$. This is generally known as the barometric altitude equation and is practical for quick estimates. H_0 and H_p are the respective scale heights for density and pressure, which are defined as

$$H_P = \frac{k_B T_{atm}}{M_{atm} g_{pl}}, \quad \frac{1}{H_0} = \left(\frac{1}{H_P} + \frac{1}{T} \frac{dT}{dh} \right) \tag{12}$$

where M_{atm} is the atmosphere's mean molecular mass, g_{pl} the planet's gravitational acceleration, k_B the Boltzmann constant and T_{atm} is the atmosphere's temperature. If the temperature is assumed to be constant throughout the entire atmosphere layer, the model is referred to as "isothermal" (refer to [33] for a more detailed treatment of the matter) and $H_P = H_0$.

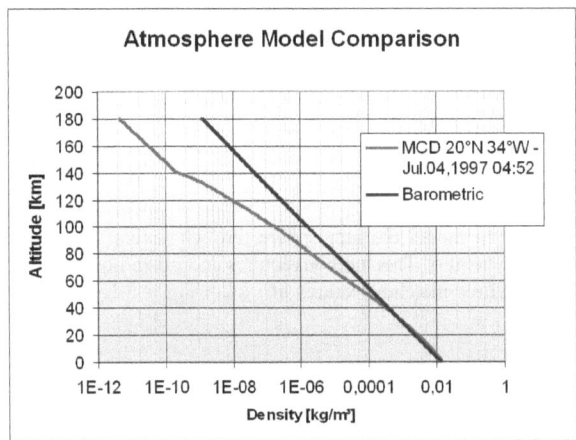

Fig. 3.3: The standard isothermal barometric atmosphere for Mars as compared to a typical profile from the Mars Climate Database.
Note that above 140 km, the MCD reverts to a standard barometric model.

Because we are mostly interested in the density profile of the atmosphere, we use the following set of equations:

$$\rho = \rho_0 \cdot e^{\frac{-h}{H_0}}, \quad p = \rho R_{Atm} T_{Atm} \tag{13}$$

where R_{atm} is the gas constant specific to the atmosphere gas mixture.

Treating the entire atmosphere of a planet as a single layer is tempting but crude, however, acceptable for lower altitudes. For Mars, current literature [34] gives the standard surface pressure as 7 hPa, the standard surface density as 0.02 kg/m^3 and the standard pressure scale height as 11.1 km. Especially at higher altitudes above 40 km though the model is rather far off the mark.

Fig. 3.3 compares this standard model to a typical profile taken from the Mars Climate Database. So for the application we have in mind – a high altitude probe – the model as such is not precise enough. However, it can be corrected for high altitudes.

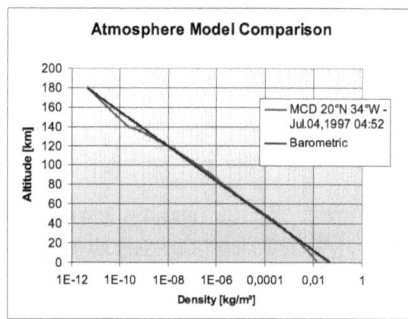 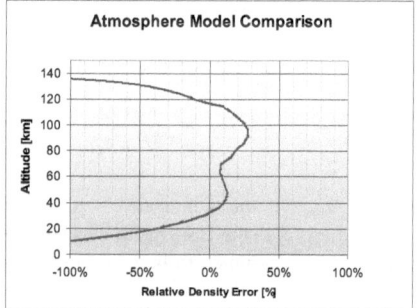

Fig. 3.4: The isothermal barometric atmosphere model density as compared to a typical profile from the Mars Climate Database. Here the surface density is 0.05 kg/m^3 and the scale height is 7770 m.

In order to use this simple model for the task at hand, we adjust the global scale height and surface conditions such that the relevant part of the atmosphere is most closely matched. For the same MCD profile as above, we select a surface density of 0.05 kg/m^3 and a density scale height of 7770m (see Fig. 3.4). This model was used in analytical solutions for parametric analysis studies, since it is accurate enough for such analyses, yet simple enough to facilitate easy mathematical treatment.

An even more accurate possibility is to model the atmosphere through several isothermal layers, all with individual density scale heights. This method requires accepting, however, that each layer has an offset in h and requires matching values of ρ_0 . For a 3-layer model, equation set (13) then expands into:

$$\rho = \begin{cases} \rho_0 \cdot e^{-\frac{h}{H_0}} & \text{for } h \leq h_1 \\ \rho_1 \cdot e^{-\frac{h-h_1}{H_1}} & \text{for } h_1 < h \leq h_2 \\ \rho_2 \cdot e^{-\frac{h-h_2}{H_2}} & \text{for } h_2 < h \end{cases} \tag{14}$$

Equation set (14) can be recomposed in terms of ρ_0 by solving set (13) for each individual layer:

$$\rho = \begin{cases} \rho_0 \cdot e^{-\frac{h}{H_0}} & for \ \ h \le h_1 \\ \rho_0 \cdot e^{-\frac{h_1}{H_0} - \frac{h-h_1}{H_1}} & for \ \ h_1 < h \le h_2 \\ \rho_0 \cdot e^{-\frac{h_1}{H_0} - \frac{h_2-h_1}{H_1} - \frac{h-h_1}{H_2}} & for \ \ h_2 < h \end{cases} \tag{15}$$

In case of a daytime Mars atmosphere profile, on latitudes neither extremely far north nor south, a good model would be $h_0 = 0km$, $h_1 = 30km$ and $h_2 = 110km$. A three layer profile with $H_0 = 11.1 \ km$, $H_1 = 8.1 \ km$ and $H_2 = 6.0km$ and a surface density of 0.015 kg/m³ is given in Fig. 3.5.

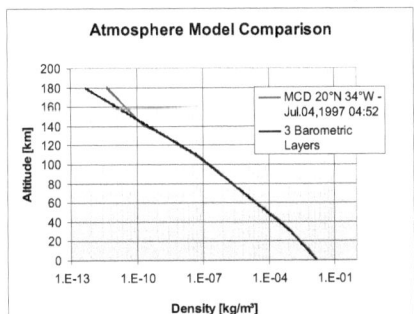 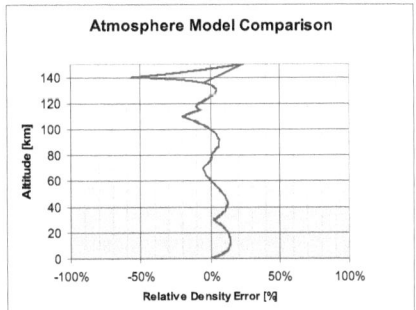

Fig. 3.5: Comparison of a three layer barometric atmosphere model against a Mars Climate Database profile to which is was matched.

This model is suitable for simple numerical trajectory integrations if neither the MCD, the GRAMM (General Reference Atmosphere for Mars Missions) nor other directly measured data tables are available, but a higher degree of accuracy is desired than would be available with a simple isothermal model.

For ballute missions into the Earth's atmosphere, such as the space flight test MIRIAM (see chapter 10.2) or the Sky Raft orbital recovery system (see chapter 10.4.2), barometric models were fitted to CIRA profiles which are typical for April at 65° northern latitude. Examples of a simple single-layer fit and a 3-layer fit with layer boundaries in 7 and 90 km are depicted in Fig. 3.6 and Fig. 3.7.

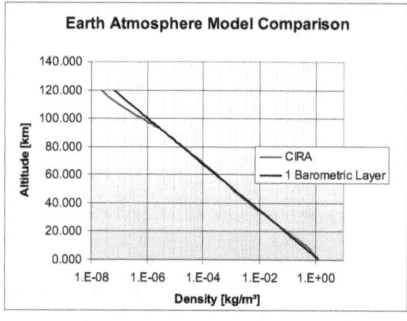 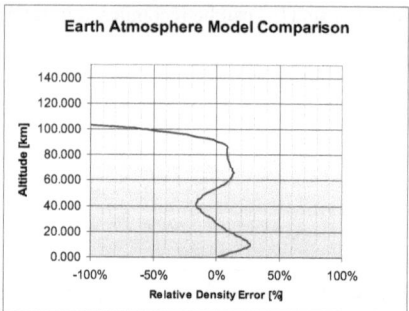

Fig. 3.6: Simple barometric atmosphere model for Earth with a density scale height of 7.2 km and a surface density and pressure of 1.326 kg/m³ and 1024 hPa fitted to a CIRA daytime profile for April at 65°North.

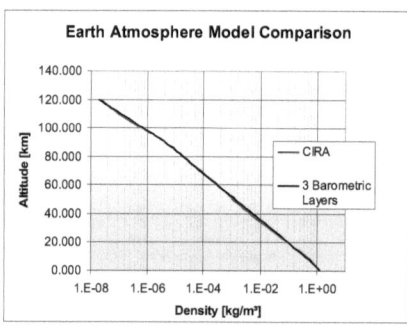 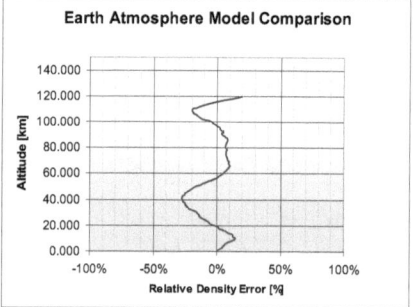

Fig. 3.7: 3 Layer barometric atmosphere model for Earth with layer boundaries in 7 km and 90 km fitted to a CIRA daytime profile for April at 65°North.

3.4   The Ballistic Coefficient and Its Impact on Mission Design

The ballistic trajectory of any object passing through the atmosphere of a planet is governed mainly by the instantaneous orientation and magnitude of its velocity vector and its ballistic coefficient. The ballistic coefficient sets into relation the mass of the object and the parameters governing its aerodynamic deceleration [27]. Therefore, this coefficient is an important parameter of a vehicle intended to enter a planetary atmosphere from space. The ballistic coefficient is defined as

$$\beta = \frac{m_{sc}}{c_d A_{sc}} \quad \text{expressed in} \quad \left[\frac{kg}{m^2} \right] \quad (16)$$

where m_{sc} is the spacecraft's atmospheric entry mass, A_{sc} the drag effective face area normal to the velocity vector and c_d is the vehicle's coefficient of drag.

Since the aerodynamic drag on a body is given by

$$F_d = q_\infty c_d A_{sc} \quad ; \quad q_\infty = \frac{1}{2} \rho_\infty |\vec{v}_\infty|^2 \tag{17}$$

with q_∞ as the free stream dynamic pressure and ρ_∞ as the free stream density [27][30], it follows that the instantaneous deceleration a_d of the vehicle is given by

$$a_d = \dot{v}_{sc} = \frac{F_d}{m_{sc}} = \frac{q_\infty}{\beta} \tag{18}$$

The deceleration of a vehicle is therefore directly proportional to the free stream atmospheric density and the reciprocal ballistic coefficient.

It is interesting to observe that the maximum g-load is most notably affected by the angle at which the object enters the atmosphere (see chapter 3.5), but the atmospheric density level at which it occurs is most prominently influenced by the ballistic coefficient.

This is plausible when considering that the ballistic coefficient represents the spacecraft's own density. For a quick look into that issue, a boxy spacecraft may be assumed, of which the density might be written as:

$$\rho_{sc} = \frac{m_{sc}}{V_{sc}} = \frac{m_{sc}}{A_{sc} \cdot l_{sc}} \tag{19}$$

where V_{sc} is the vehicle's volume and l_{sc} its length. With (19), equation (16) may be rewritten as

$$\beta = \frac{l_{sc} \cdot \rho_{sc}}{c_d} \tag{20}$$

and (18) becomes

$$a_d = \frac{\rho_\infty}{\rho_{sc}} \cdot \frac{c_d v_\infty^2}{2 \cdot l_{sc}} \tag{21}$$

From (21) follows that for a given free stream velocity, the deceleration is directly proportional to the density relation between the ambient atmosphere and the spacecraft. The deceleration profile can therefore be placed at any desirable altitude range of a given planetary atmosphere, so long as the necessary ballistic coefficient is technically feasible. An approximate, but simple analytical method on how to do that is given, as part of a more general ballute theory, in chapter 7.2 starting on page 80.

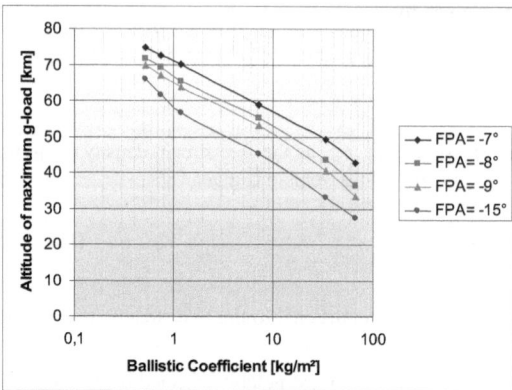

Fig. 3.8: Study of the altitude of the maximum deceleration point with respect to the ballistic coefficient for various Flight Path Angles (FPA) at entry into the Martian atmosphere.

Unfortunately no practical analytical method exists to calculate atmospheric entry trajectories in greater precision. The ones that do exist, such as [35], are cumbersome to apply and integration of numerical tables (such as a real world atmosphere profile or the dependency of the drag coefficient on the Knudsen number and flight Mach number) is problematic [36]. This makes numerical integration the method of choice. To exemplify the effects discussed above and to investigate possible mission scenarios, the entry into the Martian atmosphere from an elliptical orbit around the planet was studied assuming a perfectly spherical spacecraft, entering the atmosphere at a velocity of roughly 4.5 km/s. To characterise the altitude band in which the deceleration happens, the trajectory points of maximum deceleration and heating were chosen as a mark. Fig. 3.8 shows the altitude of the trajectory point where maximum deceleration occurs as a function of the ballistic coefficient and the flight path angle (FPA) at entry.

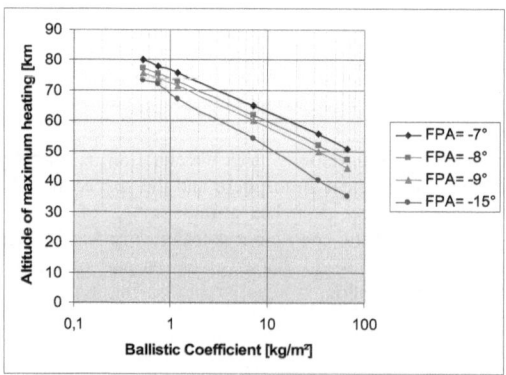

Fig. 3.9: Study of the altitude of the maximum heating point with respect to the ballistic coefficient for various Flight Path Angles (FPA) at entry into the Martian atmosphere.

With respect to scientific measurement opportunities, the altitude of maximum g-load is of course not a very practical benchmark. Instruments, however, exist that cannot function when exposed to a hot hypersonic flow. Another common goal is to maximize the measurement range and measurement time of an instrument. This is why the dwell time and altitude range, covered while operating conditions of instruments can be met, are good benchmarks for an atmospheric sounding platform. It is therefore interesting to take a look at the impact of the ballistic coefficient on the altitude at which the maximum heating rate occurs (Fig. 3.9) and how high the maximum heating rate is (Fig. 3.10), the total descent time (Fig. 3.11) and the sound barrier transition altitude (SBTA, see Fig. 3.17).

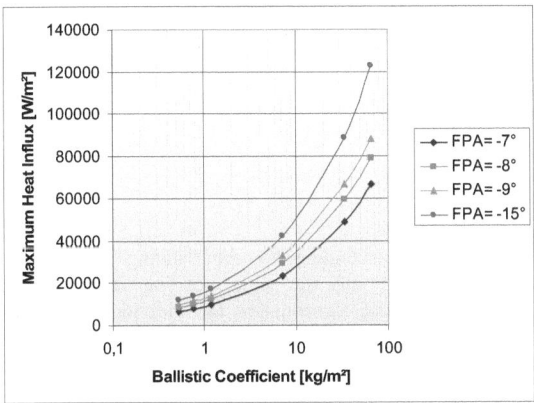

Fig. 3.10: Estimate of the maximum heat flux with respect to the ballistic coefficient using the heat flux model of Sutton & Graves [37].

As far as heating is concerned, it can be shown that for entry bodies with high ballistic coefficients the aerodynamic compression wave heating is transferred to the spacecraft wall mainly through convection and that compression wave radiation is a minor contributor [38]. Hence the maximum aerodynamic heat flux per unit area can be described according to [37] as

$$\dot{q} \propto \sqrt{\frac{\rho_\infty}{R_N}} |\vec{v}_\infty|^3 \qquad (22)$$

From (22) the impact of the free stream atmospheric density on the heat flux is readily visible, so on a vehicle with a deceleration profile high in the atmosphere (read: "with a low ballistic coefficient"), less energy from the deceleration is transferred to the spacecraft in the form of heat (see Fig. 3.10 and chapter5.3 for details). Since in (22) R_N is the vehicle's nose tip radius, blunt entry bodies heat up considerably less than sharper, pointy objects. However, blunter objects exhibit less directional stability than sharper objects, especially at speeds below Mach 3. This makes the design of an atmospheric entry vehicle with certain stability requirements (such as crewed capsules) an exercise in trade between thermal protection system materials and active attitude control system requirements (humans, for example, would suffer much inside a tumbling spacecraft).

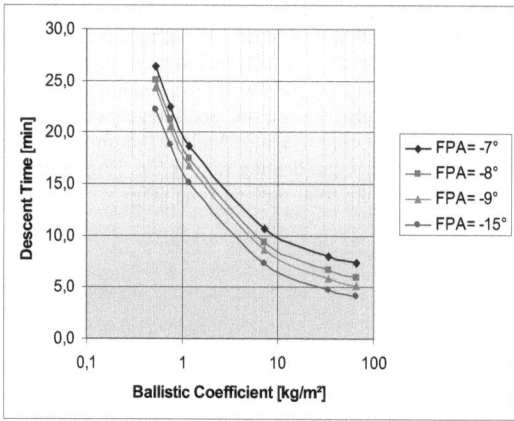

Fig. 3.11: Variation of descent time with ballistic coefficient for several entry angles.

On a probe intended for high altitude atmospheric sounding, however, landing precision and directional stability may be of lesser concern, if only the actual position can be known precisely enough to provide acceptable reference coordinates for individual scientific measurement points. In some cases, like on the Huygens Titan Probe, a roll rate may even be considered an asset, as instruments will be able to scan the surroundings of the vehicle. Assuming further that the assumptions underlying the heat flux analysis in [38] hold for lower ballistic coefficients as well, a big, spherical ballute appears advantageous.

This assumption is justified, since CFD analyses (see Chapter 5.3 starting on page 50) show that compression waves of vehicles with low ballistic coefficients exhibit no differences to conventional entries that would suggest a more opaque compression wave, or one that radiates more intensely.

3.5   Entry Angle and Entry Altitude

The magnitude of the maximum deceleration is mainly governed by the density gradient the object is subjected to along its trajectory. If the object enters the atmosphere at a shallow angle, subjecting it to a gradual increase in density, the decelerating force can be kept low. If the angle is shallow enough, ploughing through the atmosphere will not cause enough loss of kinetic energy to remain within the atmosphere altogether. Instead the vehicle will exit again and return to space, albeit on an altered trajectory. Such a manoeuvre is generally referred to as "aerobraking".

In contrast, if the entry angle is steep, the object will be subject to a sudden rise in atmospheric density, causing it to decelerate rapidly. If the angle is steep and the velocity high enough, the forces acting on the body can exceed its structural limits and the body will disintegrate.

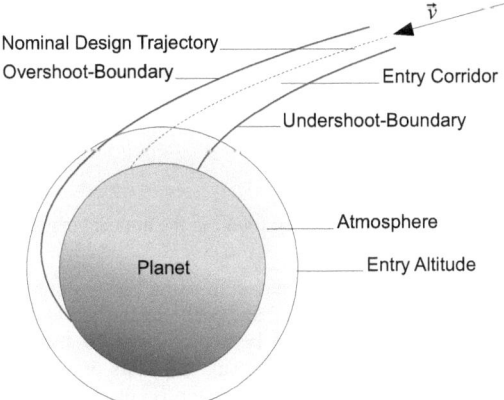

Fig. 3.12: Entry trajectory and entry corridor

The trajectories that lead to entry angles just too shallow or just too steep for the intended mission mark the boundaries of the so-called entry corridor (see Fig. 3.12). Naturally, this corridor is very mission specific.

Every atmospheric entry is characterised by the velocity and entry angle at the defined entry altitude. The entry angle γ_{entry} is the vehicle's flight path angle γ at the point of entry with respect to the local horizon (see Fig. 3.13). Since atmospheric density decreases as altitude increases, there is no discrete altitude at which the atmosphere ends and the entry altitude has to be defined otherwise.

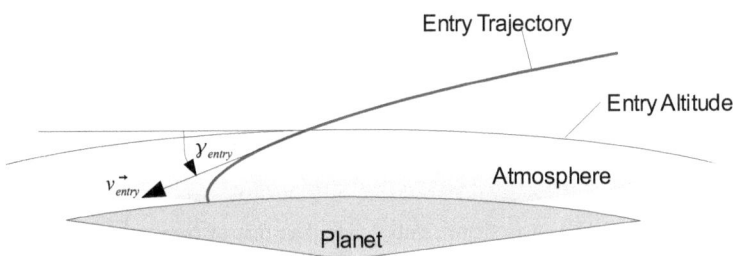

Fig. 3.13: Geometry at the point of atmospheric entry.

The entry altitude can either be fixed at a specific altitude value, or at a point where the aerodynamically induced velocity change according to (18) exceeds a fixed threshold value, $a_{d,th}$ such that the condition for entry is $a_d \geq a_{d,th}$. To do this, $|\vec{v}_\infty| = v_{entry} = v_{orbit\,@\,entry}$ may be safely assumed, as the velocities of planetary rotation and high altitude winds are at least an order of magnitude below the orbital velocity at the point of entry. Equation (18) then becomes

$$a_{d,th} = \frac{\rho_{EntryAlt} \, v_{entry}^2}{2 \cdot \beta} \qquad (23)$$

and the threshold density can be found through

$$\rho_{EntryAlt} = \frac{2 \cdot a_{d,th} \cdot \beta}{v_{entry}^2} \quad . \qquad (24)$$

With the threshold density, the pertaining entry altitude can be found in the atmosphere model of choice.

For delivering equipment to the surface of Mars, NASA has defined an entry altitude of 122km, which is practical because its spacecraft all share ballistic coefficients and velocities on the same order of magnitude. With a ballistic coefficient between 50 and 70 and entry velocities on the order of several 10^3 m/s (typical for entry capsules of Mars probes), this corresponds to an aerodynamically induced deceleration on the order of 10^{-2} m/s², a reasonable threshold considering that other perturbation forces can cause decelerations of around 10^{-2} m/s² as well. It is therefore assumed that the sensible atmosphere starts when the aerodynamic force rises notably above other perturbation accelerations.

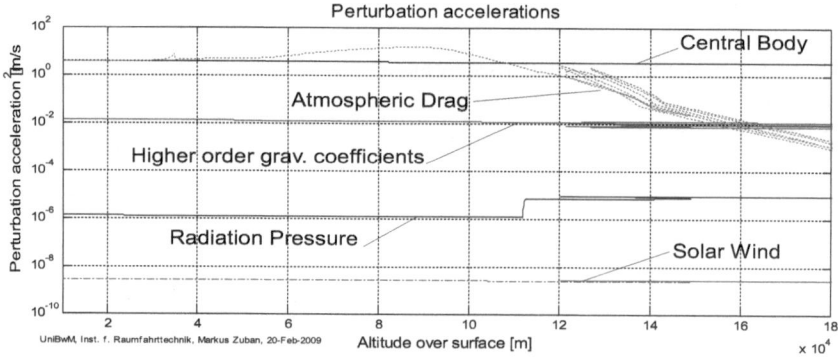

Fig. 3.14: Numerical calculation of typical acceleration forces acting on an atmospheric entry ballute with a ballistic coefficient of 0.5 kg/m² on a multiple entry aerobreaking mission.

If the ballistic coefficient is significantly lower than that of conventional entry bodies – as intended to achieve the goals set forth on the outset of this research – the entry altitude has to be raised. Fig. 3.14 shows perturbation accelerations acting on a ballute spacecraft with a ballistic coefficient of roughly 0.5 kg/m². It is clearly visible that aerodynamic drag becomes noticeable around 150 km. So when investigating the effect of a changing ballistic coefficient by comparing entry parameters, it is necessary to adapt the entry altitude to the threshold deceleration value, or else the trajectories will not be comparable. This becomes clearer when considering that the local flight path angle varies along the orbital path. A small and heavy object with a high ballistic coefficient will have a comparably low entry altitude. If a large and light object with a low ballistic coefficient enters the atmosphere along the same orbit, aerodynamic forces will have significantly altered the trajectory by the time it reaches the threshold altitude of the smaller object and hence the flight path angle will be quite different.

Since aerodynamic forces started to act earlier on the larger and lighter object, the flight path angle at this point will also be different to the one of the heavier craft at its lower atmospheric interface altitude.

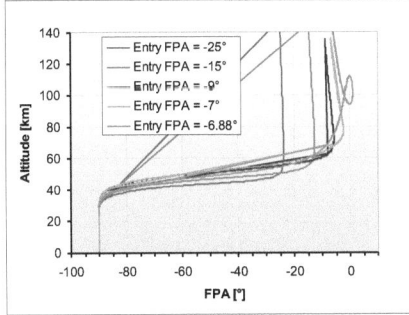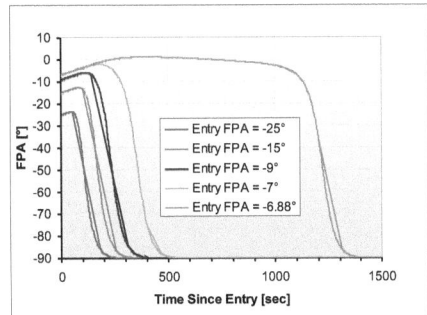

Fig. 3.15: Variation of the flight path angle over altitude and time for selected entry angles. All trajectories are numerical computations based on the Mars Climate Database and a ballistic coefficient of 0.5 kg/m².

As discussed earlier, the entry angle dictates the density gradient that the spacecraft sees. This density gradient, however, is also dependant on the atmosphere's scale height. An atmosphere with a very low density gradient (let's refer to it as a "fluffy" atmosphere) will require steeper entry angles to prevent the spacecraft from skipping back into space, whereas an atmosphere

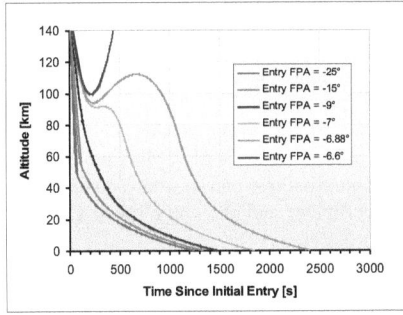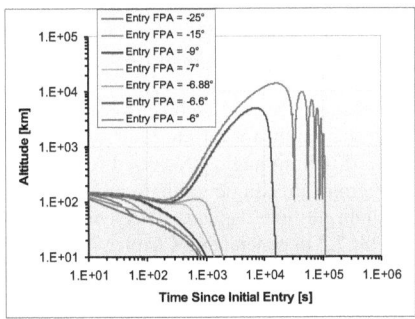

Fig. 3.16: Numerical computation of altitude vs. mission time counted from initial entry of a vehicle with a ballistic coefficient of 0.54 kg/m².

with a steep density gradient will require more shallow entry angles to prevent the spacecraft from being exposed to excessive g-loads. It is therefore interesting to note that the entry angle has a certain shallow limit which depends on the atmosphere scale height and the entry velocity and above which the spacecraft will not remain in the atmosphere any more but exit again, be it for a mere "hop" or a full revolution. Fig. 3.15 shows the variation of the flight path angle over altitude and time and Fig. 3.16 the altitude over time for typical Mars ballute parameters. If the flight path angle is positive, the spacecraft's altitude increases. We can already see that as we approach the shallow limit (in this case around -7°), a small variation in entry angle can lead to a huge effect on the trajectory. The vehicle entering at -7° takes a very long time to turn its velocity vector towards the ground. A small perturbation, perhaps caused by weather or a tiny navigational error, can therefore decide whether this vehicle is in for another orbit or a direct descent.

Therefore we can conclude that a ballute spacecraft which is intended to perform an aerobreaking mission with multiple entries requires a rather precise navigation system.

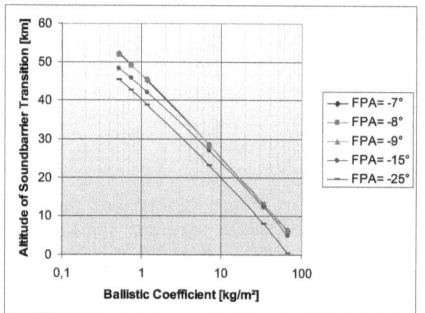 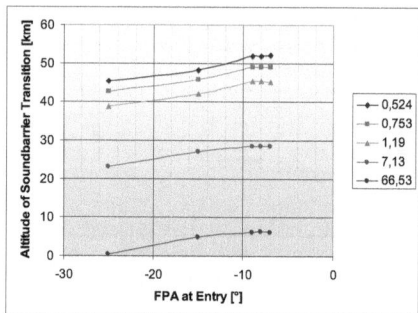

Fig. 3.17: Study of the sound barrier transition altitude for a vehicle landing on Mars with respect to its ballistic coefficient for various entry angles (left) and with respect to the entry angle for various ballistic coefficients (right).

Fig. 3.17 shows that if the flight path angle at entry is shallow enough, it does not have any significant effect on the sound barrier transition altitude any more (deviations from a smooth curve in these graphs are due to the fact that the numerical simulation is based on the Mars Climate Database, which includes inversion layers and other unsteady atmospheric phenomena). This is plausible considering that the maximum deceleration during planetary entry usually occurs at high Mach numbers. By the time the sound barrier is reached, the vehicle has already lost much of its forward velocity component from the orbit and settled into a steeper descent, subject mostly to atmospheric interaction. Therefore, an entry angle more shallow than a certain limit angle cannot raise the sound barrier transition altitude any more. This limit angle is referred to as $\gamma_{SBTA,limit}$ and in our case of a Mars atmospheric entry from an elliptic orbit is around -10°. For more shallow entries, the sound barrier transition altitude depends solely on the ballistic coefficient and the entry speed (see also Chapter 7.2 in general and Chapter 7.2.3 in particular).

It is interesting to note that if the entry is more shallow than the skipping limit (resulting in multiple aerobreaking passes, see also Chapter 3.5), which in our case study is around -7°, the sound barrier transition altitude will be lower upon final descent than it would have been with an entry angle between the skipping limit and the maximum SBTA limit (see Fig. 3.18). The reason for this is that after the aerobreaking pass, vehicle speed and entry angle have changed to result in a lower SBTA.

The maximum SBTA is therefore the upper limit of the accessible altitude range and measurement time for all equipment that needs subsonic speeds to function.

The peak is a very shallow one and therefore has little practical value. It does show, however, that the SBTA is most effectively influenced by the ballistic coefficient and that a craft may even be designed (available technologies permitting) that has its SBTA above a certain limit altitude.

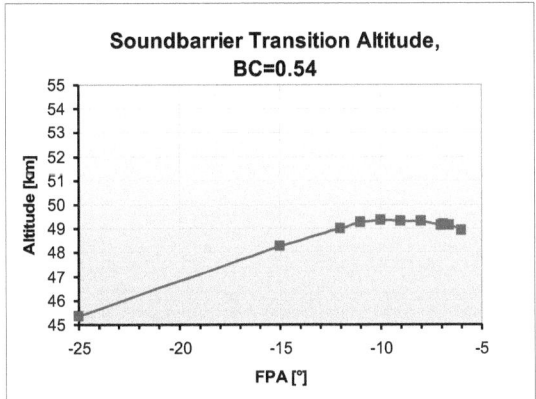

Fig. 3.18: The sound barrier transition altitude from numerical trajectory calculations for a vehicle with a ballistic coefficient of 0.54 kg/m² entering the Martian atmosphere from a 20000 x 4000 km elliptic orbit

If a higher SBTA is desirable and the technical limit for a lower ballistic coefficient has been reached, it is theoretically possible to use a lifting body ballute, such as a lentil shaped, bi-conic or a Zeppelin type ballute. These more complex concepts, however, shall not be part of this study.

4 Ballute Spacecraft Configuration Options

4.1 General Ballute Shapes and Attachment Options

Ballutes may come in various shapes and sizes, depending on their intended purpose. The configurations which were most often studied in the recent decades are given in Fig. 4.1.

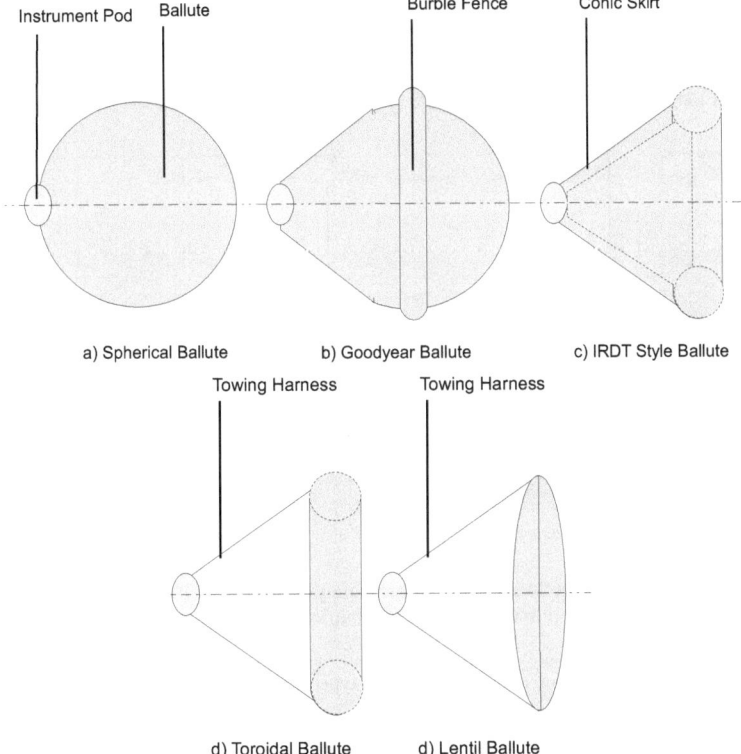

Fig. 4.1: Various ballute types and instrument pod configurations.

The sphere offers the least directional stability in flight. If left perfectly spherical, asymmetric vortex shedding will cause the vehicle to roll and tumble. This is especially so for the sound barrier transition regime. An improvement as far as directional stability is concerned would be a ballute done the Goodyear way, however, at the expense of added complexity and higher skin loads, which in turn raise material requirements considerably. The conic fore body provides improved directional stability and the addition of a burble fence prevents roll movements by ensuring equal vortex shedding. A variation of the concept would be to replace the burble fence with a string or flaps as discussed before.

To further improve directional stability, the ballute can be made completely conic, such as it was the case with IRDT (see chapter 2.4.3 starting on page 12). This ballute has a rather small interior volume and as a result also a ballistic coefficient that lies roughly an order of magnitude above that of a gossamer sphere and an order of magnitude below a classic conical entry shield. This also leads to entry heating rates that require an ablating fabric and a nose cone that can withstand a hot plasma flow.

Studied many times but never actually built are towed toroids and lentils. They offer similar aerodynamic benefits with much less hull material, but are complex mechanically, produce high skin stress and a complex aerodynamic flow field [26].

Ballute and Instrument Pod can either be directly attached to each other, or the ballute can be towed behind the Pod (see Fig. 4.2). The first configuration is referred to as a "clamped pod" or "clamped ballute", whereas the second configuration is referred to as a "towing pod" or "towed ballute".

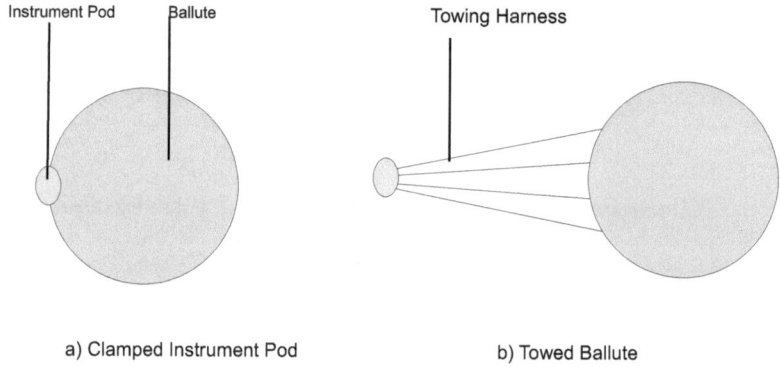

a) Clamped Instrument Pod   b) Towed Ballute

Fig. 4.2: Clamped and towed ballute configurations.

Except for the Goodyear style and the (unfortunately unsuccessful) IRDT attempts, none of these systems were ever built with the clear intention of using them as aerodynamic drag bodies.

For the prevailing application, a spherical ballute type with no additional vortex shedding was chosen.

4.2   Final Configuration

As stated, the prevailing mission objectives call for a vehicle which is able to decelerate at the highest altitude possible and subsequently descends to the ground as slowly as possible. Or in other words, this study is a quest for the lowest possible ballistic coefficient.

In terms of engineering, targeting the lowest possible ballistic coefficient is a quest for the lightest and biggest craft which is technically practical.

These requirements are best met by a big and light inflatable configuration. If indeed a mass to volume ratio (or overall spacecraft density) can be found which is on the same order of magnitude as the density of carbon dioxide, then static lift becomes a significant contributor to increasing scientific measurement time. If so desired and technically possible, a spacecraft may be designed that does not descent to the ground, but instead lingers at its equivalent density altitude like a super-pressure balloon, providing prolonged floatation time.

The emphasis on a low ballistic coefficient makes a spherical ballute the most favourable option, as it has the best surface to volume ratio and a reasonably high coefficient of drag.

Roll and tumbling are desired by some instruments. On the Huygens probe that parachuted to the surface of Saturn's moon Titan, a roll rate was artificially introduced by spinning vanes located around the perimeter of the descent vehicle. On a spherical ballute asymmetric vortex shedding provides for a sufficient roll rate without the need of additional aerodynamic measures. Since landing precision is of no concern, a purely spherical ballute appears to be the best option. No burble fence or strings need to be used to control vortex shedding for the prevailing case, but should be kept in mind as a design option in case roll rate control becomes a requirement for some instruments.

4.3   Mission Design Options

Four principal mission design options are explored in this chapter, with special emphasis on their impact on mission performance and engineering requirements. The first is a descent to a floatation altitude, the second a descent to the ground, and the third and fourth are the same as before except that they include an aerobreaking phase with several atmospheric breaking passes before their final entry.

Entry Conditions	
Entry Altitude	155 km
Entry Flight Path Angle	-10.0 °
Entry Velocity	4570 m/s

Table 4.1: Standard entry conditions for mission concept comparison.

To provide for comparable results, common entry conditions listed inTable 4.1 were used in all of the mission concept studies. The entry altitude was chosen to be far enough above the threshold deceleration of 10^{-2} m/s^2 as defined in chapter 3.5. The entry flight path angle was chosen for a shallow entry mission that does not perform several aerobreaking passes. The entry velocity was calculated assuming a de-orbit from a highly elliptical starting orbit, which is likely for the ARCHIMEDES mission if it is to fly with AMSAT P5-A (see chapter 10.3).

4.3.1   Descending to a Flotation Altitude

This concept is depicted in Fig. 4.3 and offers the prospect of a multi-day floatation mission, which maximizes the dwell time at a distinct altitude after the ballute has turned into an ordinary high altitude super pressure balloon.

Ballute / Aerobot Mass (kg)	
Instrument Pod (with 10% Margin)	4.0
Balloon Envelope	11.3
Floatation Gas	0.9
Total	16.1

Mission Mass Penalty (kg)	
Aerobot	16.1
TPS-Blanket	24.6
Aft Structure	5.7
Nose Cone Assembly	5.3
Balloon Container	0.8
Inflationsystem and Tank	7.8
De-Orbit Propellant	8.2
Margin (without Gondola)	12.9
Total with Margin	81.4

Table 4.2: Minimum Flight System Mass Budget for a Floating Ballute using optimistic figures.

To be able to float with a meaningful payload, the ballute has to be very large and very light weight, making the use of commonly available thin film materials impossible.

Therefore, the possibility was studied to make the ballute from a thinner, lighter film and wrap it in a thicker, more sturdy thermal protection blanket (TPS), manufactured from a stronger material such as NOMEX. That would allow the ballute to tolerate higher heating and mechanical loads and also ad mechanical stability in space. Such a blanket would have to be shed upon discarding the nose cover assembly as soon as an altitude regime is reached where the differential pressure between the filling gas and the environment is low enough for the balloon envelope to sustain the load. This was assumed to be the case at an altitude of approximately 10 km.

Freed from the TPS the aerobot would slowly descend towards its operating altitude, thereby taking measurements and obtaining a height dependant profile of all atmospheric data and magnetometer readings.

Fig. 4.4 gives an overview of the studied configuration. The instrument pod would be part of a gondola, suspended by a bridle that is long enough to prevent the gondola's solar cells from being shaded by the enormous balloon too frequently.

Fig. 4.3: Deployment and Entry sequence of a ballute intended for sustained floatation.

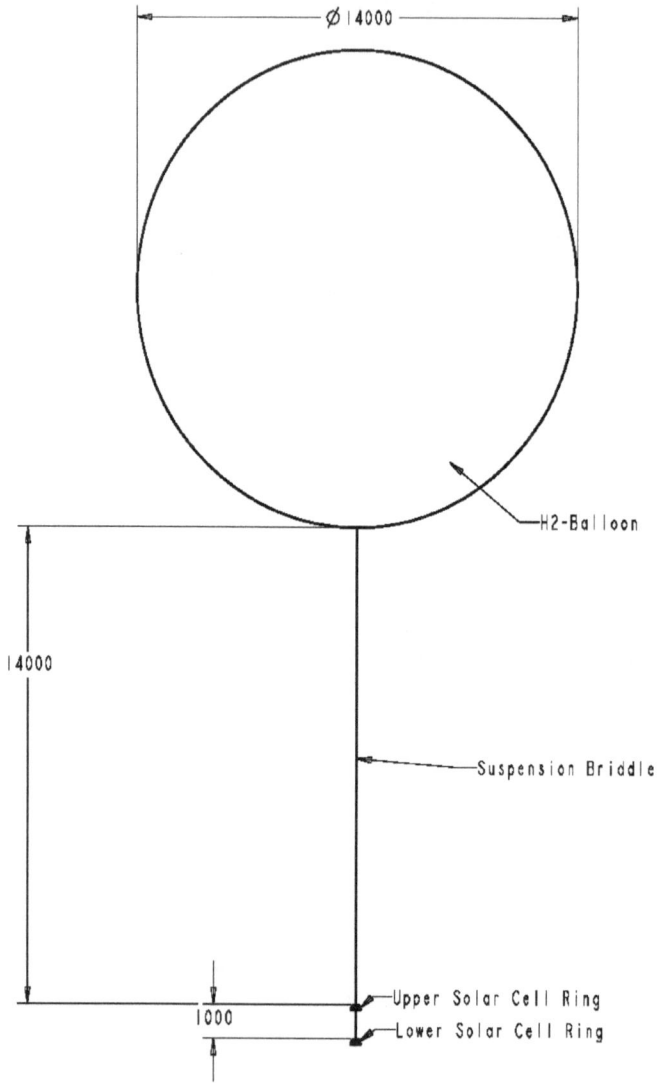

Fig. 4.4: Configuration of a possible Mars ballute that carries an instrument pod (e-box)...

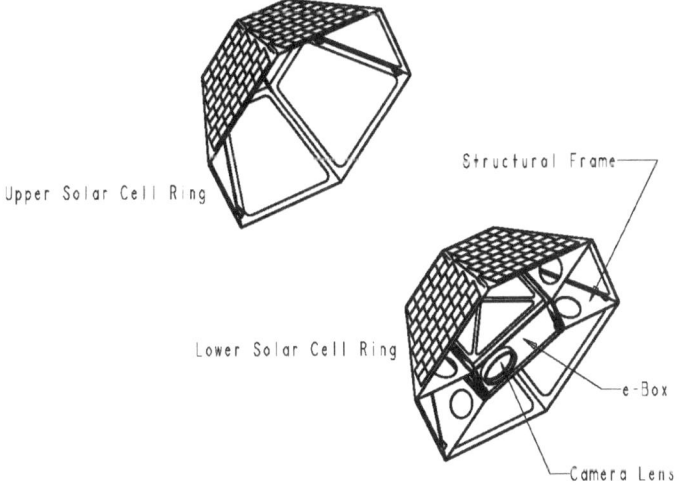

Upper Solar Cell Ring

Structural Frame

Lower Solar Cell Ring

e-Box

Camera Lens

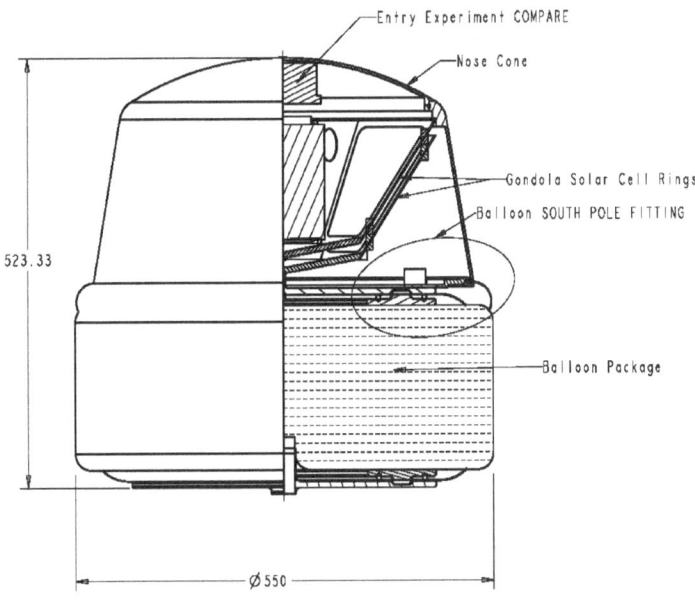

Entry Experiment COMPARE

Nose Cone

Gondola Solar Cell Rings

Balloon SOUTH POLE FITTING

Balloon Package

523.33

Ø 550

...suspended underneath the ballute's body for a multi day flotation mission.

A multi day floatation mission is certainly a very desirable prospect, purely scientifically speaking. The extension in dwell time, however, comes at the cost of added complexity in terms of solar arrays, rechargeable batteries along with recharging electronics, a very long bridle including its release cable drum and a thermal design that will have to let the tiny instrument pod survive the frigid night time temperatures on Mars. Especially the latter requirement is no easy feat. Making such a small instrument pod survive a night is close to impossible without a radioactive heating unit (RHU) at one's disposal. But such a device is not easily available outside of the USA.

But if survival of the night is technically not practical, the added complexity of a floatation mission is not warranted.

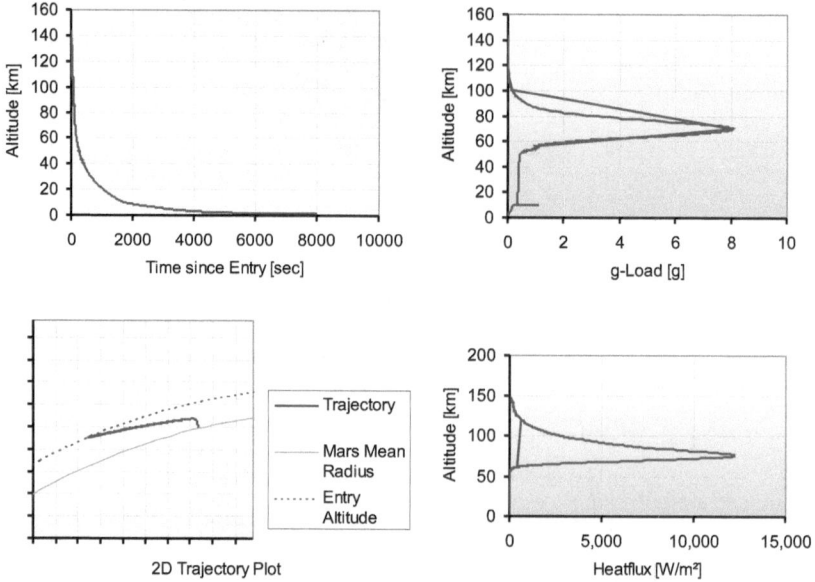

Fig. 4.5: Numerical trajectory computation of a ballute descending to a sustained floatation altitude of 2km, based on the Mars Climate Database. Areal densities are 17g/m² for a Mylar composite balloon and 40g/m² for a light TPS blanket .

To make a balloon which is capable of sustained floatation survive atmospheric entry requires a thermal protection blanket that initially drives the entry mass up by several tens of kilograms.Table 4.2 gives one of the most optimistic mass estimates conceivable, even using Hydrogen as an inflation gas, although it remains questionable whether such low masses and the transportation of the required amount of hydrogen to Mars are possible. In this table, the term "mass penalty" refers to the additional mass with which the parent spacecraft is burdened if it was to take a ballute spacecraft as outlined herein along for the ride.

Results of a numerical computation of such an optimistic mission scenario is given in Fig. 4.5. It shows that after around 10h, the floatation altitude will be reached, but heating loads well above 12kW/m² and a deceleration load of over 8 Earth g's have to be negotiated.

If the TPS blanket and the added weight of the solar cells and bridle are estimated less optimistically, they raise the ballistic coefficient, in turn raising the entry heating and making an even heavier blanket necessary. So the mass eventually rises several hundred kilograms until a possible configuration emerges. Shedding it is also not easy, since it requires the blanket to have at least two seams which can be torn open but seal tight during the hypersonic part of the flight. Plus, a device or mechanism is required that reliably tears the seam open.

So this configuration not only lowers some figures of merit (such as the total mass to science mass ratio) below limits which make the entire undertaking very questionable, it also increases the ballistic coefficient by at least an order of magnitude. While still low as compared to classic capsules, it would significantly compromise the altitude range at which the instruments could take readings (see chapter 3.4).

In conclusion, a one day floatation mission is not deemed very desirable with the current levels of technology (too much effort for too little merit) and any extended floatation mission seems not practical under reasonable mission constraints. The very long descent time alone, however, makes such a mission scientifically interesting. One possible solution could be to replace the Mylar composite and TPS blanket with a rare material referred to as PBO (see Chapter 6). Another would be to replace the suspended instrument pod / balloon configuration with a pod-less ballute. Without batteries but just solar cells, this floating ballute would not even have to survive the night, since avionics would stop functioning at dusk and automatically resume operations once the sunlight heats the film and provides electrical power at dawn.

Both concept improvements require much more fundamental research work before they can be discussed in earnest.

4.3.2   Descending to the Surface of Mars

Descending to the surface of Mars with a ballute can be desirable if scientific mission requirements call for a complete altitude coverage right down to the ground, or more simply if they do not specifically require prolonged floatation. Alternatively, if we consider the ballute as a method of delivering payloads to the surface, it might come in handy that the high-altitude deceleration profile facilitates landing at sites of higher site elevation, which would otherwise be inaccessible to heavier, faster probes. Therefore, descending to the surface of a specific minimum altitude or descending within a certain minimum time is a requirement to which a ballute offers an attractive solution.

In this concept, the instrument pod can again either be clamped to the ballute or be towing it. As far as structural engineering and practical ballute building is concerned, the advantage of this concept is that the ballute body can be heavier and smaller, allowing the use of readily available skin materials and achievable seaming requirements to arrive at a lower yet more realistic mass estimate (seeTable 4.3).

Ballute Spacecraft Mass (kg)	
Pod (with 10% Margin)	5.1
Ballute Envelope	15.1
Nose Cover Assembly	8.2
Inflation Gas	0.8
Total	**29.3**

Mission Mass Penalty (kg)	
Ballute Spacecraft	21.1
Ballute Container	0.8
Inflation System and Tank	18.0
De-Orbit Propellant	8.3
Margin (without Pod)	**10.3**
Total with Margin	**66.7**

Table 4.3: Minimum flight system mass budget for a ballute designed to descend to the surface.

This ballute spacecraft was assumed to have a PBO-reinforced Phase V ballute made of UPILEX-25RN (see chapter 6) of around 37 g/m² with a production factor of 0.3 and is inflated with Helium. It has a ballistic coefficient of 0.54 kg/m². Note that in this table, we assume that the parent spacecraft which is burdened with the additional mass penalty doubles as a service spacecraft. If this is not the case and a separate service spacecraft is required, the additional mass penalty must include the service spacecraft's structure and release mechanisms. Note also that the mass penalty can further increase if additional transponders or other equipment is needed that is only there to serve the ballute spacecraft.

4.3.3   Aerobreaking before Descending Deeper into the Atmosphere

In this concept, the ballute grazes the outer atmosphere layers repeatedly before loosing enough energy to finally descend deeper into the atmosphere (see Fig. 4.6). The process is referred to as "aerobreaking". The advantage of this method is that more measurement time becomes available in the outer atmosphere and deceleration loads are lower. However, the entry corridor is much narrower. As a result, navigational requirements are much higher and the knowledge of prevailing atmospheric conditions has to be much more precise.

The steep entry limit is obviously an entry angle leading to just two atmospheric passes, or a scientific minimum requirement. The shallow entry limit is given by available battery power and ballute envelope durability, as each atmospheric pass will subject the ballute hull material to the highly abrasive hypersonic flow field as well as mechanical and thermal load cycles.

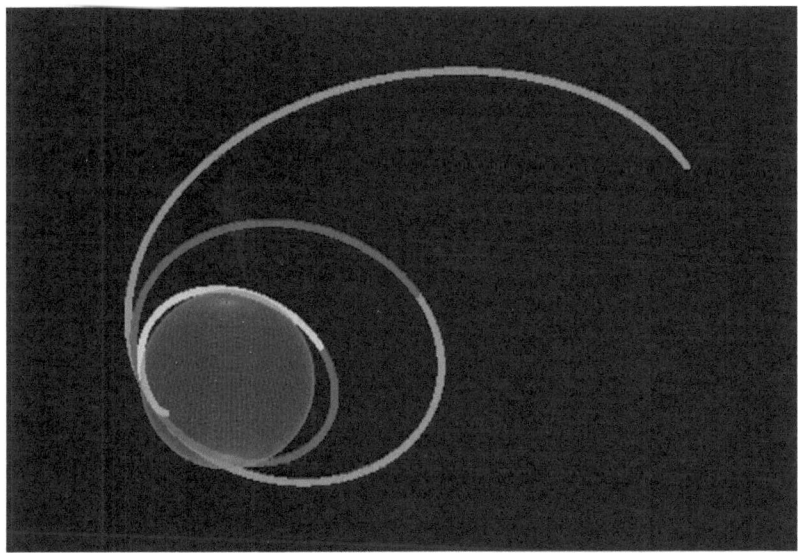

Fig. 4.6: 3D-trajectory plot of a typical aerobreaking mission. The different colours represent different orbit numbers. Numbering starts at the apocentre.

The mass budget of a ballute spacecraft is not directly influenced by such a mission design, for as long as battery life-time is of no concern. For a mission with many atmospheric passes, the available battery capacity must be increased.

5 Flight Dynamics Analysis

5.1 Overview

All research activities related to flight dynamics, aerodynamics and thermodynamics are grouped under the ARCHYFLOW sub-program, which stands for ARCHIMEDES Hypersonic Flight and Low speed descent. ARCHYFLOW analyses were performed starting with 3D trajectory simulations [39], including steady state [40] and transient [41] thermal analyses. Critical points along the hypersonic flight trajectory were further studied in in detail using CFD analyses [42][43]. Eventually, aeroelastic effects were investigated by performing modal and deformation analyses [44][45].

Trajectory simulations were made using numerical and analytical methods of various complexities and precisions. The most concise and sophisticated tool used is a full 6-DOF trajectory integration based on a powerful radio science mission simulator [46] developed at the Institute of Space Technology of the University of the federal Armed Forces of Germany (UBW) [47] and is presented in Chapter 5.2. Aerodynamic heating was estimated in all trajectory calculations using the simplified analytical method of Sutton & Graves [37] (see also chapter 7.2.1.4 on page 91). The European Mars Climate Database (MCD) was used as a global atmosphere model [32].

For good measure it should be noted that precise analytical solutions exist that also offer good results. Especially references [35] and [36] report on an optimized analytical model to study ballute trajectories and aerobreaking missions. None of these were used in this analysis, however, as their application is not easier than numerical integration. Since a good and validated numerical method was already available, using it was deemed more practical for the task at hand.

Simple analytical methods suitable for parametric analyses were developed and are presented in chapter 7.2. These are not intended to replace more precise analyses, but are tailored to suit the needs of trade studies and primary system design iterations.

Note that this chapter is an introduction to the applied methods with only few example results. Mission specific analysis results can be found in the respective chapters treating particular mission scenarios.

5.2 Mission and Sensitivity Analysis with the Radio Science Simulator

Originally conceived to simulate interplanetary missions with an emphasis on preparing and analysing radio sounding experiments, the Radio Science Simulator (RSS) is a highly versatile and very precise tool for numerical trajectory analysis. It already provides a full six degree of freedom orbit integration including perturbation forces as subtle as solar radiation pressure, the gravity of other celestial bodies and a host of other minuscule forces. It also comes with a gravity potential model of Mars of the highest known degree and the aforementioned Mars Climate Database as a reference atmosphere. In addition to this, the location of planets and any point on their surface can be calculated. Another great advantage of the RSS is that the calculations of vehicle speeds and locations were routinely proven to be in excellent agreement with real world radio sounding observations done with the space probes Mars Express and Venus Express.

5.2.1   Atmospheric Entry Trajectories

For atmospheric entry trajectory simulations with the RSS, atmospheric drag was added to the set of perturbation forces [39].

The variation of the drag coefficient with Mach number (see Fig. 5.1) was directly taken from wind tunnel testing results obtained by the Goodyear Corporation in the 1960s [48]. Even though the shape of the studied ballute system differs somewhat from the perfectly spherical ballute suggested herein for hypersonic entry, the effect of this difference on the Mach-number dependency of the drag coefficient is considered small as compared to other uncertainties. Especially as reference flight data for such a configuration does not presently exist anyway.

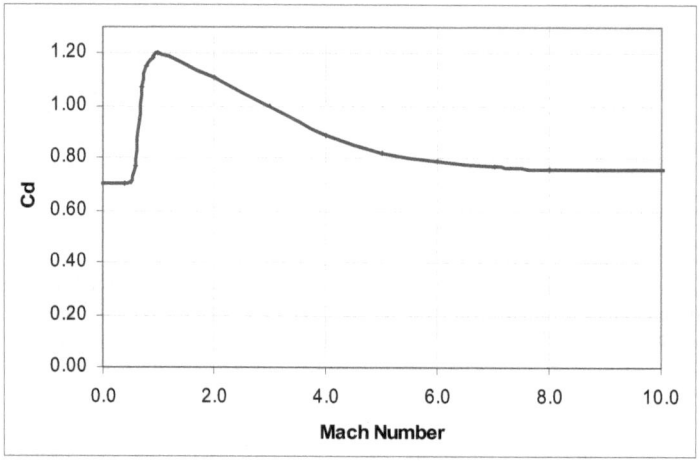

Fig. 5.1: The variation of the coefficient of drag with flight Mach number for the Goodyear ballute (based on data obtained by [48]).

The orientation of the spacecraft was assumed to be always pod-forward, as the instrument pod pulls the CG to a location outside the geometric centre and therefore should result in a pod-forward attitude during atmospheric entry. For the simplicity of the model, the same attitude was assumed for trajectory parts outside the atmosphere. Since the deformation of the ballute during hypersonic flight is expected to be minimal, the departure of the ballute from a perfectly spherical shape was not modelled. Hence, aerodynamic lift was also not modelled.

In contrast, static lift was introduced as a perturbation force. An atmospheric sounding ballute may be designed to have significant aerostatic lift to prolong descent- and measurement time in the lower atmosphere.

Selected plots of a typical result of such an analysis are shown in Fig. 5.2 (in this case for an ARCHIMEDES orbit version designated Alpha-1 Norm with two atmospheric entries). The analysis starts with the de-orbit manoeuvre. The 3D orbit plot gives an overall picture of the orbit in planetocentric coordinates and the ground track plot shows the trajectory ground track of the ballute across the actual surface. The deceleration plot clearly shows the first and second atmospheric entry deceleration peaks and the point at which the nose cover assembly is discarded.

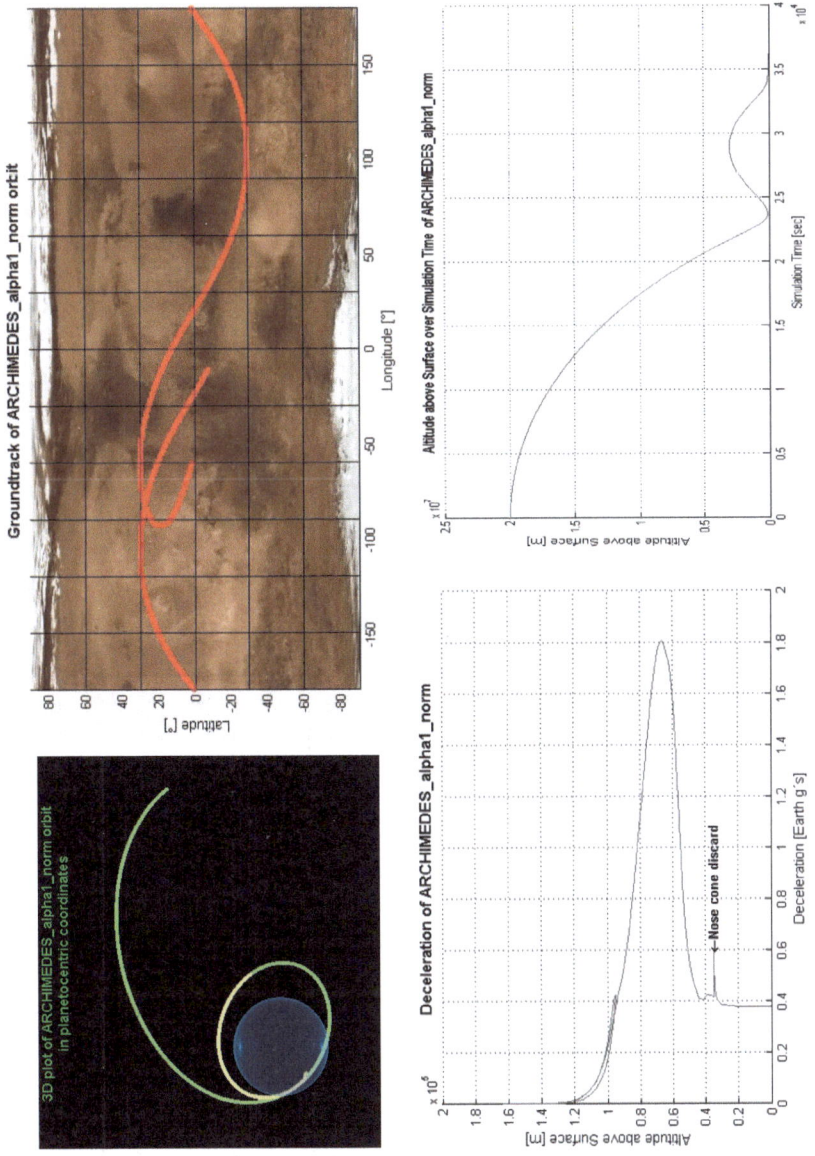

Fig. 5.2: Selected results of a typical ARCHIMEDES trajectory analysis (in this case the Alpha-1 Norm orbit).

5.2.2   Mission Design Analysis

Because of its concise nature, the RSS is also ideally suited to study the sensitivity of ballute missions and to analyse mission geometries. For the ARCHIMEDES ballute spacecraft, it is necessary to ensure that atmospheric passes do not occur during eclipse and that they are visible from the telecommunications orbiter to ensure a permanent telemetry stream during that critical mission phase.

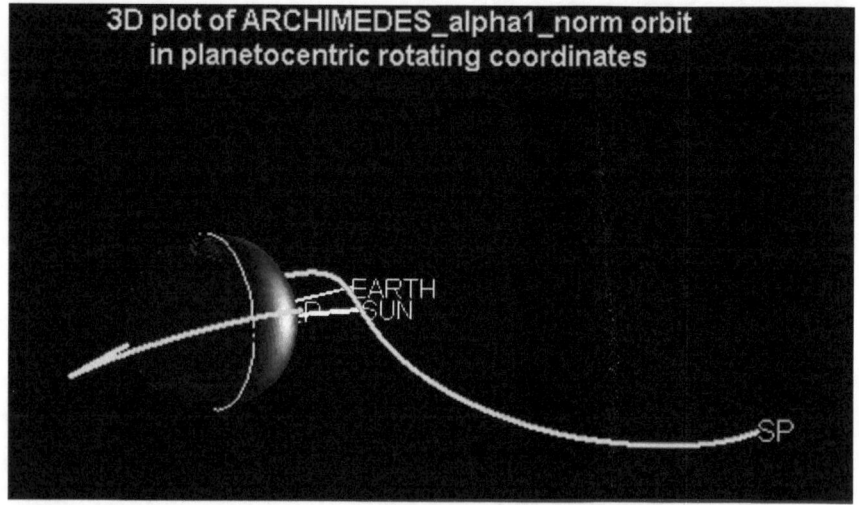

Fig. 5.3: Illumination analysis at the time of the ballute spacecraft's touch down on Mars. The yellow line indicates the terminator.

To do this, a tool was added to the RSS that can calculate the range and visibility between two spacecraft and perform a radio link analysis [49]. Two individual RSS simulations are loaded simultaneously and their individual range, visibility, possible data rate, lighting conditions and their visibility from Earth is then calculated and plotted. For project ARCHIMEDES orbits of the AMSAT P5-A Mars satellite and the ARCHIMEDES ballute were analysed that way, starting from the point of ARCHIMEDES' de-orbit.

A typical illumination analysis for the ARCHIMEDES Alpha-1 Norm type orbit is given in Fig. 5.3. The depicted situation is the time of touch down. Note that the Earth and Sun are visible at that time.

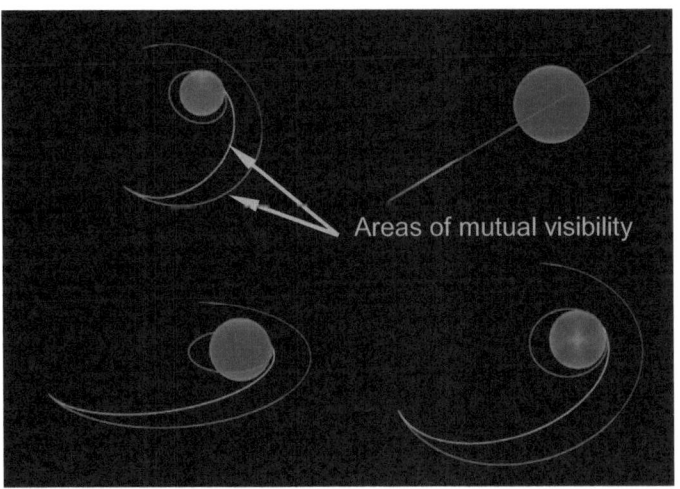

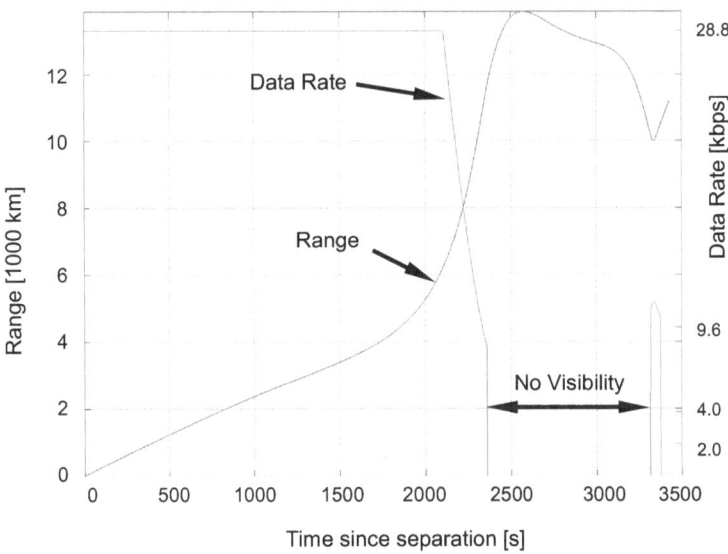

Fig. 5.4: Data rate analysis between the orbiter and the ballute for a typical ARCHIMEDES mission. The right ordinate shows available modes only. (adapted from [49]).

Results of the ballute to orbiter visibility and link analysis are given in Fig. 5.4. Bright trajectory colours indicate mutual visibility between the two spacecraft. Note that the atmospheric passes are visible, but only barely and that there is only a little coverage margin. This scenario has therefore still room for improvement. The data rate in Fig. 5.4 was calculated based on the ARCHIMEDES telemetry subsystem as given in 10.3.3.2 on page 218. The most notable feature is the assumption that only omnidirectional antennas with 0dB gain are available on either spacecraft. This is a worst case assumption because the orbiter has to track Earth with its high gain reflector antenna and is not assumed to have a separate antenna on a gimbal. The ballute spacecraft is assumed to have an uncontrolled attitude and anyway too little room to house anything more complex than a simple dipole integrated into the ballute envelope. Data rate calculations further assume the theoretically possible maximum under a predefined link margin (in this case 6dB) and a predefined coding scheme. A possible maximum data rate that the orbiter can handle was also defined. These graphs can now be used for the telecommunication subsystem engineering process to design the best data rate switching and power output strategy for an optimal communication strategy.

5.3   Aerothermodynamics and Aeroelasticity

5.3.1   Computational Fluid Dynamics Analysis of Critical Trajectory Points

In the framework of the ARCHYFLOW programme detailed aerothermodynamic analyses were performed to improve the knowledge about the thermal and mechanical loads and the dynamic behaviour of the balloon.

Two reference points were selected for the analyses: the point of the trajectory where the maximum stagnation point temperature occurs, determined as the radiation adiabatic temperature by a model based on [37] and the trajectory point of maximum deceleration.

For Mars entry, the analysis was performed by [42] based on a modification of the hypersonic CFD code CEVCATS-N for a rarefied high-altitude hypersonic flow field with chemical non-equilibrium [50].

CEVCATS-N is a Navier-Stokes solver, originally developed for low-speed continuum flows

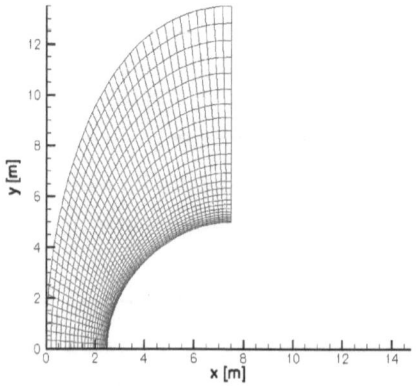

Fig. 5.5: Grid around the ballute for ARCHIMEDES CFD Analyses.

by the DLR Institute of Aerodynamics and Flow Control. It is based on a structured mesh storing flow variables at cell vertices (the mesh used for this analysis is given in Fig. 5.5 and typical results for the point of maximum heating in the standard entry trajectory is given in Fig. 5.6). Adapting this code for hypersonic entry requires the important assumptions that the shock front is transparent to thermal radiation, which has already been confirmed for a continuum flow and entry velocities much higher than for the prevailing case [38].

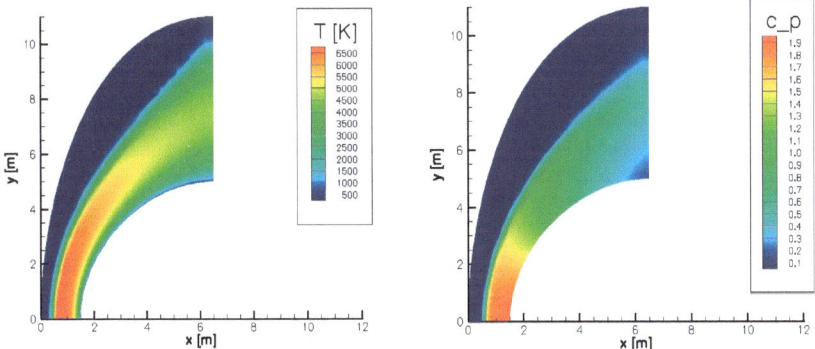

Fig. 5.6: CFD results for flow-field temperature and pressure coefficient of an ARCHIMEDES type ballute spacecraft at the point of maximum heating rate as determined with CEVCATS-N [42].

Despite the high altitude of the deceleration profile and the low density of the Martian atmosphere there, the application of a Navier-Stokes solver is justifiable because of the big diameter of the ballute. Navier-Stokes equations are a mathematical model of a continuum flow [28] and are not applicable to a rarefied particle flow, where atmospheric molecules are so far apart that they hardly interact with each other. In order to determine if enough molecules hit the passing space vehicle to justify the assumption of a continuum flow, the dimensionless number known as the Knudsen number [27] can be evaluated. It sets into relation the mean free path length of the atmosphere molecules and the characteristic length of the spacecraft:

$$Kn = \frac{\lambda}{L} \tag{25}$$

where λ is the mean free molecular path length and L is the characteristic length of the spacecraft. As the spacecraft plunges through the atmosphere, seeing a continuously changing ambient density, the Knudsen number changes continuously as well. If this relation is significantly larger than 1, mostly free molecular flow persists, in which molecules act like independent particles. If the Knudsen number is significantly below one (the limit is generally considered to be 10^{-2} [27]), we can speak of a continuum flow and rarefaction effects can be neglected. Anything in between is considered a transitional flow regime, which exhibits properties of both the continuum flow and the rarefied molecular flow.

The mean free path length of the molecules is the statistical mean distance a gas particle has to travel until it collides with another one. It can be calculated with the following relation [27]:

$$\lambda = \frac{1}{n\,A_{molec}\sqrt{2}} \tag{26}$$

where n is the number of particles per unit volume and A_{molec} the effective face area of a gas molecule. This can be recomposed in terms of more readily available numbers [51][52] by replacing n and A_{molec} :

$$\lambda = \frac{k_B\,T}{p}\cdot\frac{1}{\pi\,d_{molec}^2\sqrt{2}} \tag{27}$$

with T and p as the atmosphere temperature and pressure and d_{molec} as the approximate molecular diameter. Around the points of maximum deceleration and maximum heating rare, the Knudsen number for a ballute spacecraft configured like ARCHIMEDES, with a ballistic coefficient of 0.5 kg/m², entering a daylight Martian atmosphere is on the order of 10^{-3} (see Fig. 5.7) and even less inside or behind the shock wave where the gas gets compressed.

To estimate which type of flow field (rarefied, transitional or continuum) may be expected for a given ballute design along its trajectory, we seek an altitude-dependant relationship between the Knudsen number and the ballistic coefficient. Note that for this thesis we limit our analysis to spherical ballute spacecraft. We replace the number of molecules per unit volume with an expression for the molecular mass and the atmospheric density ρ [52]:

$$n = \frac{\rho}{m_{molec}} \tag{28}$$

We then replace the density with our altitude-matched exponential atmosphere model from equation (13). The Knudsen number from equation (25) is then expressed in terms of the altitude h by:

$$Kn = \frac{m_{molec}}{\rho_0\,e^{\frac{-h}{H_0}}}\cdot\frac{1}{A_{molec}\sqrt{2}}\cdot\frac{1}{d_B} \tag{29}$$

Note that the relevant characteristic flow field dimension is the ballute diameter d_B .

Because the points of maximum deceleration and heating rate are of particular importance, knowledge about the flow field there is required in order to choose the appropriate flow field simulation method. In order to express the Knudsen number at the altitude of maximum deceleration $h_{a,max}$ and the altitude of maximum convective heating $h_{c,max}$ in terms of the ballistic coefficient β , we use an analytically derived relation for the two altitudes as a function of β (see chapters 7.2.1.3 and 7.2.1.4 for details):

$$h_{a,max} = H_0\cdot\ln\left(\frac{H_0\rho_0}{\beta\sin\gamma}\right)$$

$$h_{c,max} = H_0\cdot\ln\left(\frac{3H_0\rho_0}{\beta\sin\gamma}\right) \tag{30}$$

In equation set (30), H_0 is the density scale height, ρ_0 the atmosphere pressure on the surface, β the ballistic coefficient and γ the entry angle. With equation set (30), equation (29) can be expressed as a function of the ballistic coefficient such that

$$Kn_{a,max} = \frac{H_0}{\beta \sin \gamma} \cdot \frac{m_{molec}}{A_{molec} \sqrt{2}} \cdot \frac{1}{d_B} \; ; \quad Kn_{c,max} = \frac{3 \cdot H_0}{\beta \sin \gamma} \cdot \frac{m_{molec}}{A_{molec} \sqrt{2}} \cdot \frac{1}{d_B} \tag{31}$$

Finally, we solve the ballistic coefficient equation (16) for the diameter of a spherical ballute so that

$$d_B = \sqrt{4 \frac{m_{sc}}{c_d \pi \beta}} \tag{32}$$

and (29) finally becomes

$$Kn_{a,max} = \frac{H_0}{2 \sin \gamma \sqrt{\frac{m_{sc} \beta}{c_d \pi}}} \cdot \frac{m_{molec}}{A_{molec} \sqrt{2}} \; ; \quad Kn_{c,max} = 3 \cdot Kn_{a,max} \tag{33}$$

This equation shows that the more shallow the entry angle γ is or the lower the ballistic coefficient, the higher the Knudsen number will be. If the spacecraft mass rises but the ballistic coefficient must remain within design limits, the Knudsen number decreases because the diameter of the ballute (the characteristic flow field dimension) has to increase as well.

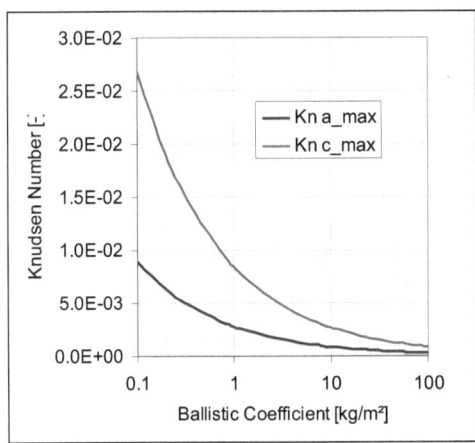

Fig. 5.7: The Knudsen number for the trajectory points of maximum heating rate and maximum deceleration as a function of the ballistic coefficient for a 36 kg spherical ballute spacecraft.

Fig. 5.7 shows the Knudsen number for the points of maximum heating and maximum deceleration of ballute spacecraft like the ones studied herein, entering the Martian atmosphere at -10°. The graph shows that all practical ballutes with realistic ballistic coefficients can be regarded as being in a continuum flow during these two critical mission phases and can therefore be treated with Navier-Stokes solvers for CFD analysis.

Dust effects within the flow were not considered, as we may conclude from limb sounding experiments with the HRSC camera on Mars Express [53] that dust particles in the atmosphere do not reach altitudes at which the main deceleration profile of a configuration such as ARCHIMEDES occurs.

Because of symmetry assumptions, a 2D quarter-circle was used as a base shape, with the grid around it being more detailed close to the surface (see Fig. 5.5).

Results obtained with CEVCATS-N for a typical ARCHIMEDES Alpha-1 Norm type entry trajectory as calculated with the RSS and a simple atmosphere model composed of a 97% CO_2 and 3% N_2 are given in Fig. 5.6, where flow field temperatures and pressure coefficients are shown for the point of maximum heating.

Despite the huge and light body entering the atmosphere of Mars, chemical nonequilibrium is still a factor. For the same trajectory point and vehicle configuration the concentration of the main flow field constituents is shown in Fig. 5.8.

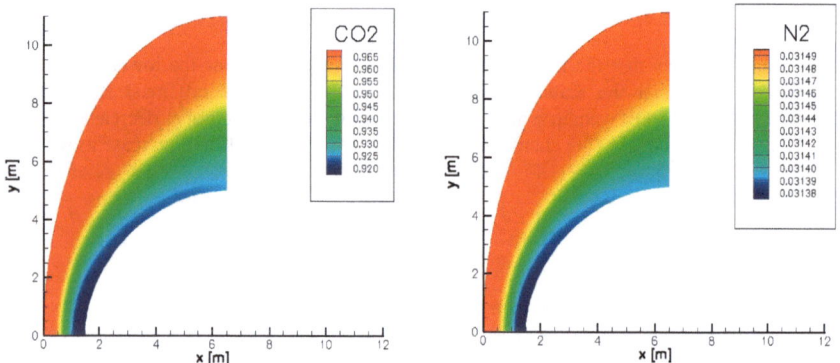

Fig. 5.8: Concentration of Mars' main atmosphere constituents (carbon dioxide and nitrogen) around the ballute spacecraft at the point of maximum heating rate as determined with CEVCATS-N [42].

5.3.2   FEM Stress and Modal Analyses

Two different finite element (FEM) analyses using MSC Nastran were made for an ARCHIMEDES-type 10-m ballute and both were used to analyse ballute envelope behaviour at the two critical trajectory points of maximum heating and maximum deceleration. The first model was a simple sphere with no detail whatsoever [44], that provided a quick look and an estimate of the general ballute envelope behaviour.

Based on experience gained with this model, a more sophisticated model was then made for an in- depth analysis [45]. The inertia relief method was used for this, which allows the simulation of unconstrained structures (such as flying vehicles) in a static analysis. Instead of fixing it at dedicated points, the structure is constrained by its own moment of inertia.

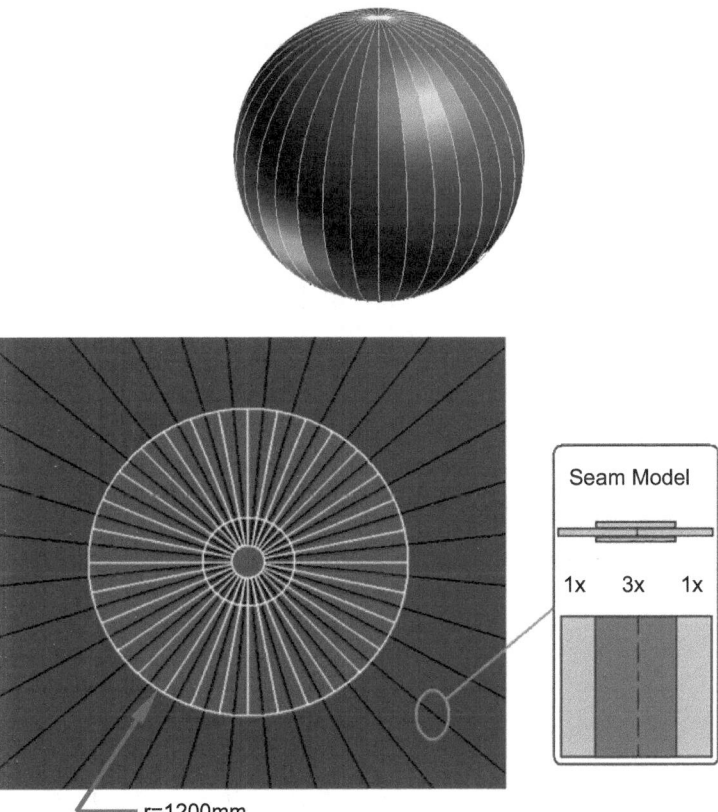

Fig. 5.9: FEM model of the ballute envelope (from [45]).

The ballute envelope was modelled assuming a beach ball pattern, as presented in chapter 7.3.2, with 32 longitudinal elements, seams and polar cap reinforcements, extending 1200mm outward from either pole (see Fig. 5.9). The skin material properties modelled were those of UPILEX®-RN as described in chapter 6. Envelope elements and seams were assumed to be cylindrical sections. The seam and polar cap elements were modelled with three times the thickness of the skin element, to represent a sandwich seam as outlined in chapter 7.3.4. Any other details like adhesive layers, envelope instrumentation, antennas and wires were omitted. The instrument pod was modelled as a single massive element, with a diameter of 600mm and a mass of 15 kg.

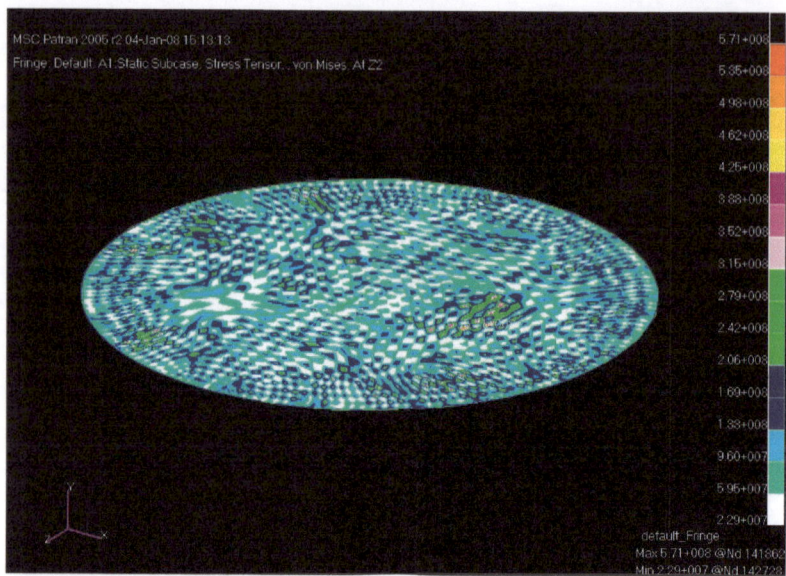

Fig. 5.10: Static stress field in the north polar cap of a ten metre ballute.

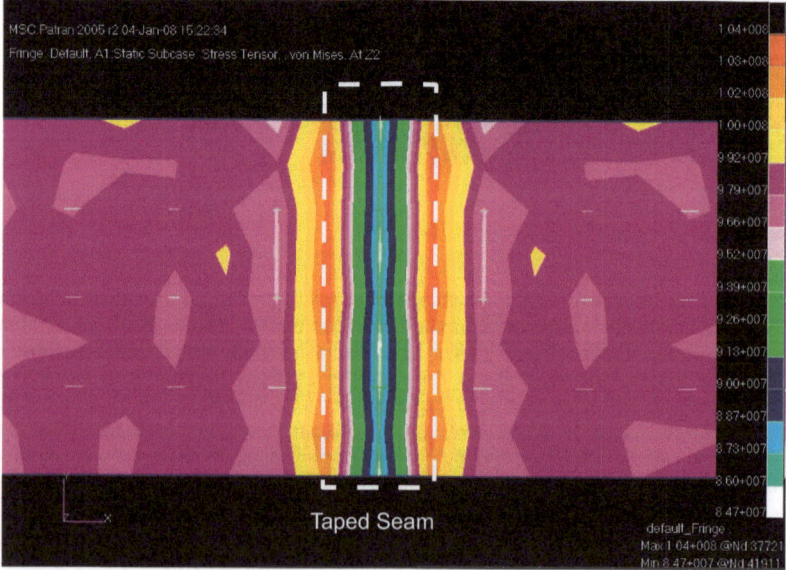

Fig. 5.11: Typical stress field across the seam of a 10-m ballute inflated with 10hPa nominal pressure at the point of maximum deceleration. Maximum load is 104 Mpa [45].

The north polar cap presents a particular challenge in terms of discretisation. In contrast to the south pole, where the instrument pod is mounted and modelled as a rigid body, the north polar cap is part of the ballute skin and must therefore also be discretisised. The challenge is that the elements get smaller and slimmer towards the actual pole, where triangles seem inevitable. A good FE model, however, should have similarly sized and proportioned elements all over. The problem was solved by using a particular computer generated and seemingly chaotic mesh that avoids the use of triangles and maintains similar dimensions for all elements. Fig. 5.10, which shows a static stress map of the north polar cap on a 10-metre ballute, also clearly shows the mesh. Note the large stress differences in various adjacent elements, which is a typical problem with FE models. The analytical solution is an average of these results, so the FEM result in this picture is interpreted accordingly. Note that the FE analysis of the north polar cap does generally yield results that show a large fluctuation in load which is not realistic. It is only carried along to maintain a homogeneous distribution of element dimensions and sizes throughout the entire model and to avoid triangular elements.

For the stress analysis, the forces which were modelled are the interior gas pressure, atmospheric static pressure, flow field dynamic pressure and the inertia of the instrument pod. As before, an instrument pod forward attitude was assumed in all cases. The dynamic pressure was taken from the CFD analysis, dividing the southern hemisphere in 48 equally wide ring segments with incident angles from 90° (pole) to 0° (equator).

The FEM stress analysis is in good agreement with the simpler analytical methods presented in chapter 7.2.2 and is therefore not considered to be critical. Fig. 5.11 shows the stress distribution field across a double-splice same-width tape seam, whereby it should be noted that the slight jagged appearance is solely due to the imperfect FEM mesh discretisation. Note the stress peaks of 104MPa at the fringe of the seam, which is where the material will likely fail first. That's why it is two individually wide tapes should be used, which help reduce this peak (see chapter 7.3.2).

As expected from pressure difference considerations (see also chapter 7.2.2 starting on page 96), relative pod displacement at the point of maximum deceleration lies below 0.5 % of the total diameter.

For the modal analysis, only a simple sphere was modelled including the filling gas. Nastran's Normal Mode analysis was used, which neglects all attenuation effects. The first eigenmode for a 10-metre ballute was found around 41Hz and the second around 50Hz. Increasing the envelope skin thickness by a factor of 6 decreased the first eigenfrequency to 39Hz and the second to 47Hz. The skin thickness is therefore found to be negligible and reinforcements and seams can safely be omitted in such an analysis. Reducing the diameter on the other hand does have a significant effect. A ballute of 5 metres was found to have its first natural eigenfrequency at 80Hz, and the second at 96Hz.

The eigenfrequency of the inflation gas bubble also has an effect. According to this analysis the inflation gas pressure is influencing the eigenfreuqency as much as several Hertz depending on temperature. Fig. 5.12 shows the first and second eigenmodes of the ballute.

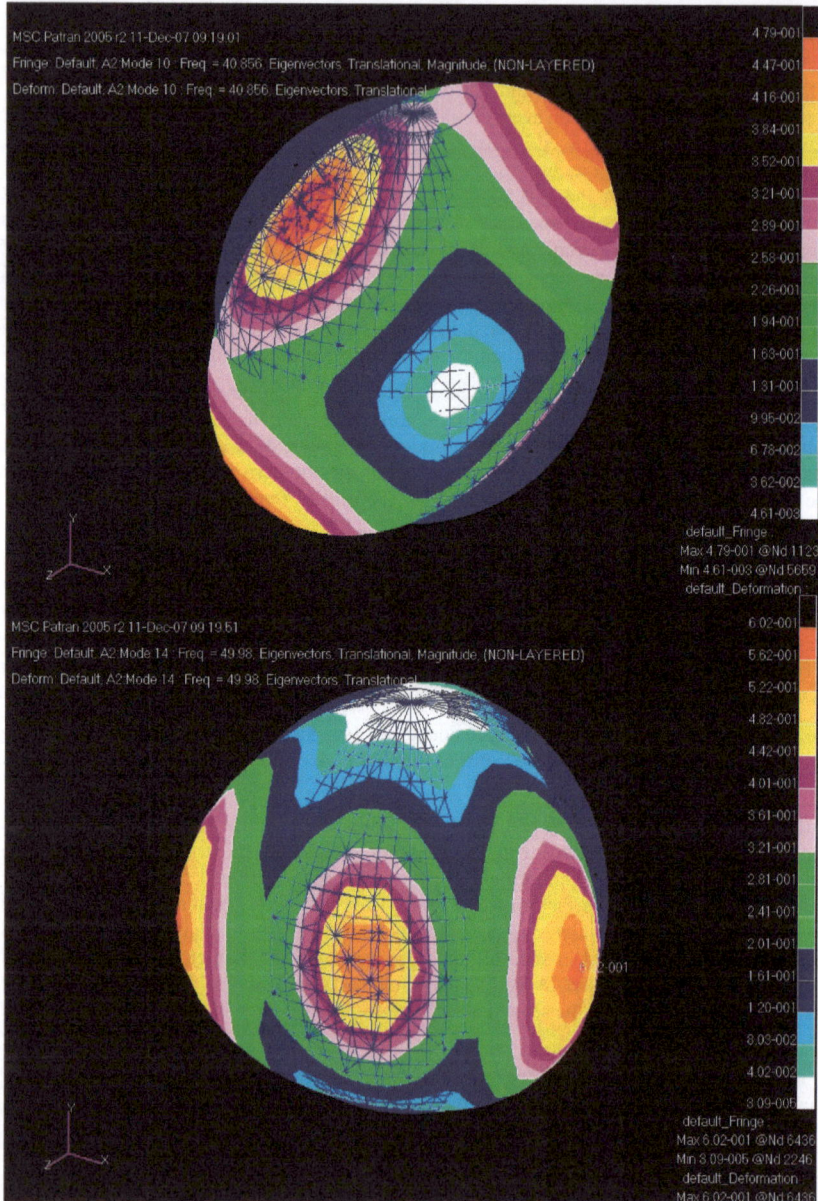

Fig. 5.12: First (top) and second (bottom) eigenmode of the spherical thin-film ballute.

5.4   Thermal Analysis

Like any other balloon, a ballute requires a thorough thermodynamic analysis. Its internal pressure, descent characteristics and life expectancy depend largely on its thermal properties. To analyse the thermal behaviour of the ballute during all mission phases (in space, during hypersonic flight and final descent), a simple yet effective thermal node-model based on [41] was created and included in the numerical mission evaluation computation. Heat sources are solar and planetary radiation, latent heat from the instrument pod and hypersonic compression wave heating. Heat sinks are the space environment and the planetary atmosphere.

5.4.1   Overview and Coordinate System

The rectangular coordinate system (Fig. 5.13) is centred inside the ballute and is body fixed, oriented such that the X-axis points through the south pole, where the instrument pod lies, assuming that this also coincides with the flight direction (the velocity vector coincides with +X). Because the ballute is a rotational symmetric object, the direction of the other two axes is not fixed by special features on the ballute, but the model grid (see Fig. 5.13).

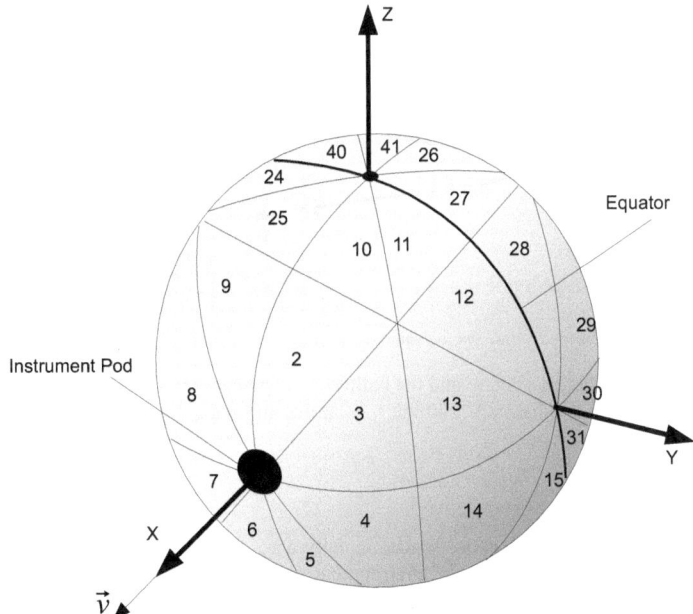

Fig. 5.13: Sketch of node model for thermal analysis, showing the coordinate system and the distribution of nodes across the ballute surface.

Fig. 5.14 shows the definition of angles and directions. The two vectors $\vec{n}_S$ and $\vec{n}_{PL}$ are unit vectors describing the position of the sun (subscript S) and the planet's geometric centre C_{PL} (subscript PL). Each is defined by two position angles, α and β, such that

$$\vec{n}_S = \frac{1}{\sqrt{1+\sin^2\beta_S}} \begin{pmatrix} \cos\alpha_S \\ \sin\alpha_S \\ \sin\beta_S \end{pmatrix} \tag{34}$$

and

$$\vec{n}_{PL} = \frac{1}{\sqrt{1+\sin^2\beta_{PL}}} \begin{pmatrix} \cos\alpha_{PL} \\ \sin\alpha_{PL} \\ \sin\beta_{PL} \end{pmatrix} \tag{35}$$

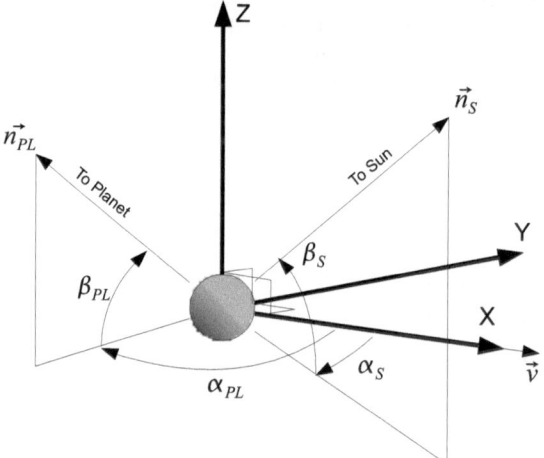

Fig. 5.14: The coordinate system used for the thermal analysis.

The distance between the planet's centre and the ballute (not shown) is further divided into the planet's radius R_{PL} and the ballute's altitude h above the planet's surface, so that the vector ballute-planet may be written as

$$\vec{C}_{PL} = (R_{PL}+H)\vec{n}_{PL} \tag{36}$$

The balloon was modelled with 51 nodes spaced equally across the surface (Fig. 5.13). The instrument pod, inflation gas and the space environment were modelled with one node each. The instrument pod was modelled as a homogeneous body of aluminium. For the entire analysis, an attitude with the south pole forward (instrument pod pointing in flight direction) is assumed.

5.4.2   Basic Equations

The thermal model [41] is based on the first law of thermodynamics [54], which states that the energy is conserved in any process. The internal energy U of a system can only change if heat Q is added or work W is done so

$$dU = \delta Q - \delta W \tag{37}$$

In the node model, this equation has to be satisfied by every vertice, so that we may form a node-equation:

$$W_i \frac{dT_i}{dt} = \sum_j C_{ij}\left[T_j - T_i\right] + \sigma \sum_j R_{ij}\left[T_j^4 - T_i^4\right] + P_i + Q_i \tag{38}$$

where W_i is the heat capacity of the node, T_i its temperature, C_{ij} and R_{ij} are the heat conductivity and radiation exchange factor between node i and node j, P_i is the heat dissipation from the node (i.e. the instrument pod), Q_i is the heat flowing into or out of the node and σ is the Stefan-Bolzmann constant.

This set of equations is solved for all nodes of the model. Reference [41] has more details on the mathematical background and how it was programmed.

5.4.3   Radiative Heat Exchange

Because the ballute's thermal behaviour is based almost exclusively on radiative heat exchange, we shall afford a closer look. Radiative heat influx comes mainly from the sun, from the planet's own infra-red radiation, and through solar radiation reflected off the planet's surface (albedo). Heat exchange inside the ballute is also mostly radiative, as thermal conductivities are very low, and ballute surface elements all see each other plus the instrument pod. Heat also comes from aerodynamic friction and the compression wave during entry. It is mainly radiated off into deep space.

To model the radiative heat exchange, a line of sight $\vec{x}_s$ from the node to each heat source must be determined. This line of sight is defined as

$$\vec{x}_S = \lambda \vec{n}_S \quad \text{and} \quad \vec{x}_{PL} = \lambda \vec{n}_{PL} \tag{39}$$

using the unit vectors defined in chapter 5.4.1 and λ as the distance to the heat source. If λ is positive, the resulting vector points towards the heat source, if λ is negative, it points towards the ballute.

To determine whether the ballute is lit by the sun or eclipsed by the planet, the terminator plane is introduced. The terminator plane is defined such that the the terminator circle (the day/night boundary) lies within it (see Fig. 5.15).

All planets of the solar system are far enough away from the sun to safely assume that on a planetary scale, the solar rays are parallel to each other. Hence, the surface normal of the terminator plane coincides with with $\vec{n}_s$.

For the purpose of this analysis, it is also safe to assume that heat influx variations by atmospheric scatter and a partially obstructed sun are negligibly small. So the shadow zone is assumed to extend cylindrically behind the night side of the planet, parallel to $\vec{n}_s$. For the same reason penumbra is neglected so the planet's geometric centre lies within the terminator plane and the plane is defined through

$$\vec{\epsilon}_{TP} \Rightarrow \vec{n}_S\left(\vec{x} - \vec{C}_{PL}\right) = 0 \tag{40}$$

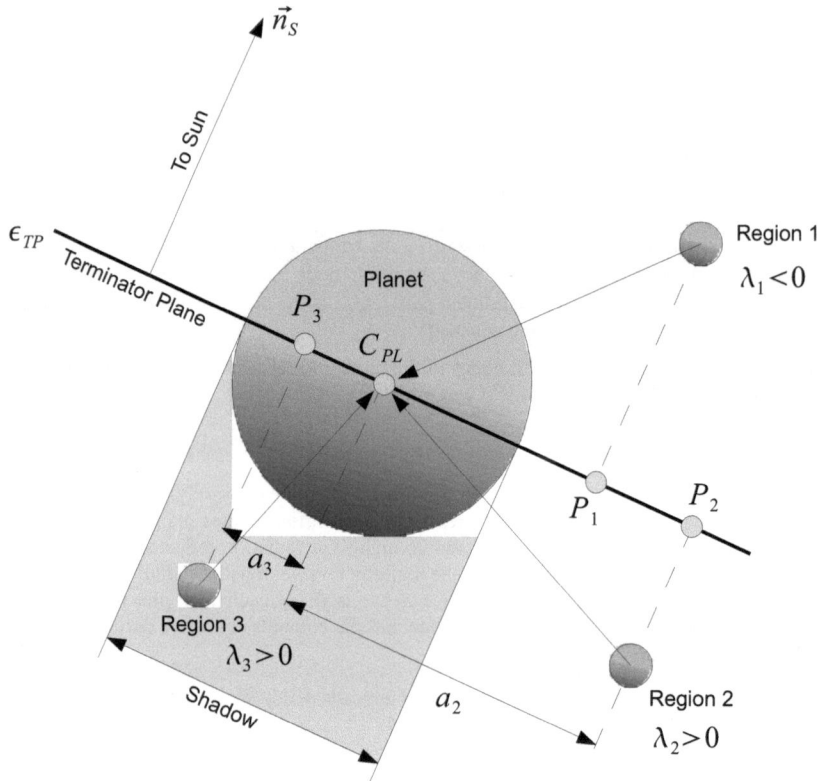

Fig. 5.15: Possible ballute positions relative to the planet.

The ballute can either be above the terminator plane (region 1) or below it (region 2). If the ballute is below this plane, it can also be eclipsed by the planet (region 3). To decide which region the ballute is in, the point P_P , at which the sun vector passes through the terminator plane, has to be found. The distance and direction to the terminator plane can then be found by solving (39) and (40) for a corresponding λ_P . If λ_P is positive, the resulting vector points towards the sun and thus the terminator plane lies between the ballute and the sun. In this case, the distance a_P between the planet's centre and P_P has to be found. If a_P is smaller or equal to R_{PL} , the ballute is eclipsed.

Without proof, using the previously defined unit vectors to solve (39) and (40) for λ_P yields

$$\lambda_P = \frac{2\,(H+R_{PL})(\cos[\alpha_{PL}-\alpha_S]+\sin[\beta_{PL}]\sin[\beta_S])}{(3-\cos[2\,\beta_S])\sqrt{1+\sin[\beta_{PL}]^2}} \tag{41}$$

The distance a_P is quite simply determined by

$$a_P = |\vec{C}_{PL} - \vec{P}_P| \tag{42}$$

Which heat sources have to be taken into account depends on the region the ballute is in. Using subscript P as a region index, the three regions may be defined as follows

Region 1: $\lambda_1 \leq 0$
If the ballute is in this region, heat influx from the sun, the planetary albedo and the planet's infrared radiation have to be taken into account.

Region 2: $\lambda_2 > 0$, $a_2 > R_{PL}$
In this region, only solar and planetary radiation are considered. The determination of albedo from a crescent planet would require a much more complex mathematical analysis, making a numerical view factor determination program necessary. Given the nature and purpose of the analysis, as well as its inclusion in a numerical trajectory computation for basic mission evaluation, this is considered both unnecessary and impractical.

Region 3: $\lambda_3 > 0$, $a_2 < R_{PL}$
Only planetary radiation has to be considered in this region, as the ballute is eclipsed.

The heat influx Φ in $[W]$ on a given surface element ΔA is generally described by

$$\Phi = -\oint \vec{q}\, \overrightarrow{dA} \tag{43}$$

where $\vec{q}$ is the flux density vector in $[W/m^2]$ and $\overrightarrow{dA}$ the surface element normal of ΔA. If the entire surface A is divided such that all resulting elements have the same area ΔA, an individual element j may be described by

$$\overrightarrow{dA}_j = \vec{n}_j \Delta A \tag{44}$$

with $\vec{n}_j$ as the unit vector of the surface element normal. For each surface element, the absorbed flux densities are given by

$$\vec{q}_S = -\vec{n}_S \alpha_{B,VIS}\, q_{SolarConst} \quad \text{for solar radiation,} \tag{45}$$

$$\vec{q}_{PL_i} = \frac{-\vec{n}_{PL}}{\left(1 + \dfrac{H}{R_{PL}}\right)^2} \alpha_{B,IR}\, \varepsilon_B\, \sigma\, T_{PL}^4 \quad \text{for planetary infra-red and} \tag{46}$$

$$\vec{q}_{PL_A} = \frac{-\vec{n}_{PL}}{\left(1 + \dfrac{H}{R_{PL}}\right)^2} \alpha_{B,VIS}\, a_{albedo}\, q_{SolarConst} \quad \text{for planetary albedo.} \tag{47}$$

All of the above heat flux densities are in $[W/m^2]$. The planetary infra-red radiation is simplified by σT_{PL}^4 , $q_{SolarConst}$ is the local solar constant, a_{albedo} the planet's albedo coefficient and $\alpha_{B,IR}$, $\alpha_{B,VIS}$ and ε_B the ballute material's absorption and emission coefficients, respectively.

To avoid the need for a complex numerical view factor computation, the dependence of the visible planetary surface area on the distance from it is neglected. Radiative heat exchange is calculated using view factors. A view factor is defined to be the relation of the heat radiated towards another surface element devided by its total radiated heat. The view factor between each ballute and the planet for each surface element is thus determined by

$$F_{B-PL}=\frac{\vec{n}_j\vec{n}_{PL}}{\left(1+\dfrac{H}{R_{PL}}\right)^2} \qquad (48)$$

The view factor ballute–space is quite simply obtained by subtracting F_{B-PL} from one:

$$F_{B-Space}=1-F_{B-M} \qquad (49)$$

Regardless of the region in which the ballute is in, the question remains whether the surface element under consideration sees a specific heat source or not, as it might be obstructed by the rest of the ballute. To obtain this information, the scalar product of the surface normal and the vector towards the source can be evaluated. In case of the sun, the surface element is lit if the condition

$$\vec{n}_S\cdot\vec{n}_j\geq 0 \qquad (50)$$

is satisfied. If so, the result is a factor expressing the effectively exposed area, since the scalar product of the unit vectors yields the cosine of the angle between them. Naturally, the same method may be applied using $\vec{n}_{PL}$ instead of $\vec{n}_S$ to determine if the surface element j sees the planet:

$$\vec{n}_{PL}\cdot\vec{n}_j\geq 0 \qquad (51)$$

The total absorbed power P_j per surface element can now be obtained by combining all of the above into the following set of equations:

Solar radiation:

$$P_{Sj}=\vec{n}_j\cdot\vec{n}_S\alpha_B q_{SolarConst}\Delta A \quad \text{if (50) is satisfied and}$$

$$P_{Sj}=0 \quad \text{if (50) is not satisfied.} \qquad (52)$$

Planetary infra-red radiation:

$$P_{PLj_i} = \frac{\vec{n}_j \cdot \vec{n}_{PL}}{\left(1 + \dfrac{H}{R_{PL}}\right)^2} \alpha_{B,IR} \varepsilon_{PL} \sigma T_{PL}^4 \Delta A \qquad \text{if (51) is satisfied and}$$

(53)

$$P_{PLj_i} = 0 \quad \text{if (51) is not satisfied.}$$

Planetary albedo:

$$P_{PLj_A} = \frac{\vec{n}_j \cdot \vec{n}_{PL}}{\left(1 + \dfrac{H}{R_{PL}}\right)^2} \alpha_{B,VIS} a_{albedo} q_{SolarConst} \Delta A \qquad \text{if (51) is satisfied and}$$

(54)

$$P_{PLj_A} = 0 \quad \text{if (51) is not satisfied.}$$

Radiative heat dissipation into space:

$$P_{j-Space} = \left\{ 1 - \frac{\vec{n}_j \cdot \vec{n}_{PL}}{\left(1 + \dfrac{H}{R_{PL}}\right)^2} \right\} \varepsilon_B \cdot \Delta A \cdot \sigma \cdot (T_{Space}^4 - T_j^4) \qquad \text{if (51) is satisfied and}$$

(55)

$$P_{j-Space} = \varepsilon_B \cdot \Delta A \cdot \sigma \cdot (T_{Space}^4 - T_j^4) \quad \text{if (51) is not satisfied.}$$

The view-factors for radiative heat exchange inside the ballute are quite simply obtained. by considering a hollow sphere. Because all surface node elements have the same size (as per model definition), the view factor from each element to each other element is simply $1/n$, whereby n is the total number of surface elements [40][41]. The radiative heat transfer is then calculated for each node using the current element temperature and the same thermo-optical property values as for the external radiative heat exchange. The latter assumption results from the assumption that all studied configurations relied on a single-layer ballute skin design.

5.4.4 Example Results

Results from this thermal model can be used as a basis for the material analysis. The above calculations show a temperature which is a little lower than the temperatures obtained from CFD analyses, since material heat capacities and balloon internal convection were taken into account. However, the general heat-flux distribution obtained from the CFD analyses were used in the implementation of this thermal model, which was made part of the RSS trajectory simulation. Typical results, plotted against mission time, are shown in Fig. 5.16. In this set of plots, all nodes are shown simultaneously. Note the short eclipse period during which the temperature drops as low as 150 K for all nodes, requiring active heating for the payload and avionics suite.

Another important fact is that the comparatively low skin temperatures during entry (as compared to classic probes) are also possible because of heat being radiated off the "hot" hemisphere towards the "cool" hemisphere in the vehicle's wake field.

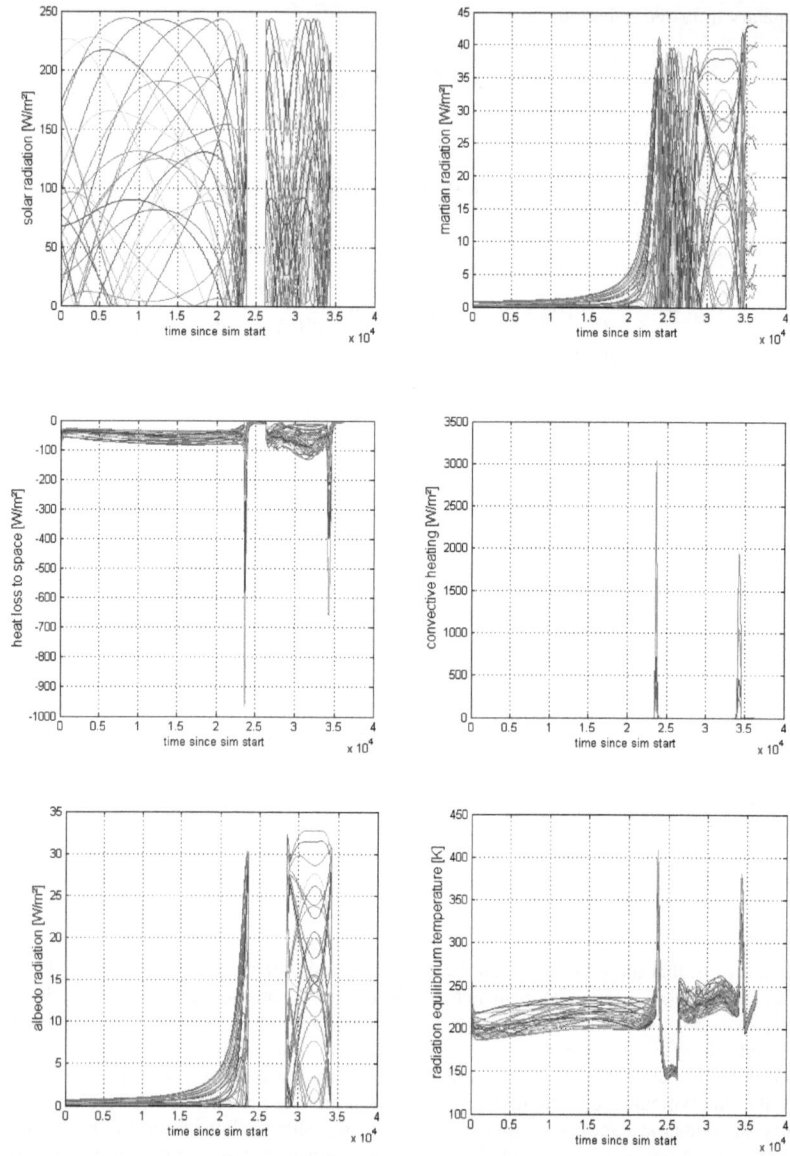

Fig. 5.16: Typical heating results for all vertices across the ballute surface for an ARCHIMEDES Alpha-1 Norm type entry trajectory.

6 Material Analysis

Finding the right material for the ballute envelope is paramount and represents a key technology. In order for the ballute to resist high pressure and temperature differences during hypersonic flight, yet be light enough for a reasonable mission, it has to be made of a high performance thin-film material or a laminate thereof. Such materials are tested as part of project ARCHIMATTER, which stands for ARCHIMEDES Material Testing and Research. It involves testing of candidate materials under expected mission conditions. Project ARCHIMATTER is also targeted at exploring seaming and production techniques, including the development and testing of high strength high temperature resistant adhesive tapes (refer to chapter 7.3.4) and a number of welding techniques.

6.1 Material Selection

The ballute envelope material or material laminate has to be flexible enough to be packed into an acceptable volume, stable enough to retain its mechanical properties while packed and light enough to fulfil mission requirements. It has to be impermeable to gas and strong enough to hold up during hypersonic flight. Note that this last point not only includes high temperatures, but also chemical and mechanical stability in a hypersonic flow field environment, that might contain chemically aggressive constituents dissolved in the shock wave (see chapter 5.3), but also an abrasive particle stream (dust). Last but not least, the material has to conform to space qualification standards, as defined in ECSS-Q-70-71A rev. 1 [55] and ECSS-Q-70-02A [56], as well as planetary protection requirements [57]. The material also has be readily available in large enough quantities to fashion a series of ballutes from it.

The only group of materials that are up to the task are polymers. Such materials exhibit a viscoelastic behaviour, which means their mechanical properties depend on the time they are subjected to a load, as well as the frequency of possible load cycles. This behaviour is caused by the fact that tension leads to the relocation and stretching of molecular chains. A combination of loads (like mechanical and thermal loads or a certain load cycle frequency) always leads to different results than the individual load alone. Therefore materials not only have to undergo testing under single load conditions (like a pure stress-strain test), but also simulated mission conditions, which combine several or all expected loads simultaneously (load cycles on top of a static biaxial load in a heated oven, or atmospheric entry testing under a real hypersonic conditions).

An initial survey of available materials and their suitability for a Mars balloon mission was made for the ARCHIMEDES project by the Institute of Statics and Dynamics of Aerospace Structures (ISD) at the University of Stuttgart, an effort that later even received ESA funding [58]. The study concluded by identifying several possible plastic thin films for surface or air deployed balloons, but that most materials fail at conditions encountered during hypersonic flight. However, studies already exist investigating balloons for the lower Venusian atmosphere, which is also a high temperature and chemically aggressive environment (see [59] and [60]).

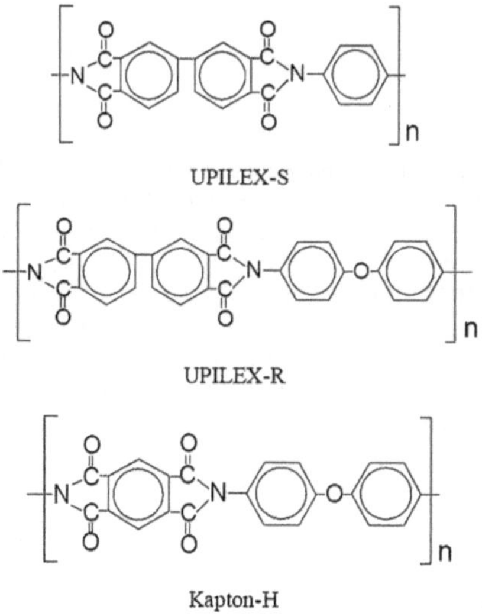

UPILEX-S

UPILEX-R

Kapton-H

Fig. 6.1: UPILEX® molecular structure compared to a common Kapton variant [63] (UBE Industries).

Based on all of these finding, two polymers named UPILEX®-S and -R have been identified as the most promising candidates. Fig. 6.1 shows their molecular structure, compared to the more widely known polyimide Kapton. The thermoplastic material UPILEX® is a polyimide which is known for good mechanical properties at high temperatures. Another material, known as Polyphenylbenzobisoxazole (PBO), promises much better performance, but is not commercially available as a thin film and appears to be sensitive to damage induced by radiation. Its molecular formula can be seen in Fig. 10.34. The company Foster-Miller of Waltham, Massachusetts, has already built subscale test balloons made of PBO for NASA [61] and was willing to deliver small film samples for testing at the UBW lab for material analysis, so the applicability of PBO for ARCHIMEDES was also investigated (see chapter 6.2.2). The molecular structure of PBO is shown in Fig. 6.2.

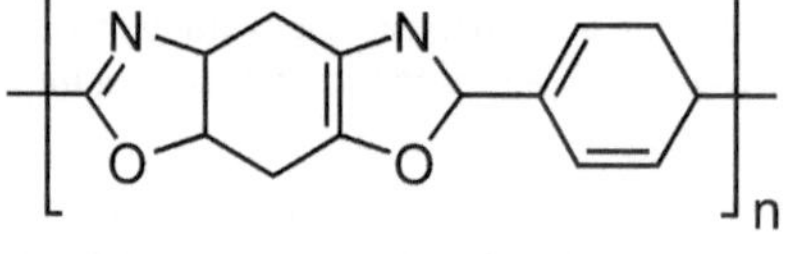

Fig. 6.2: Molecular structure of Poly(p-phenylene-2,6-benzobisoxazole), also known as PBO or its trade name Zylon (based on information from Toyobo).

Results show that even much degraded PBO offers a far superior performance than UPILEX®, but availability remains a huge problem. The UPILEX® subtype SN also comfortably exceeds mission requirements, but is almost impossible to pack, because it is stiff and easily breaks. UPILEX®-RN, in turn, boasts excellent processing and packaging properties, but mission margins are minimal.

It is therefore practical to either investigate the necessary technology to manufacture large PBO thin films of constantly good properties or to design a ballute with UPILEX®-RN, that has structural reinforcements made of PBO where appropriate (see chapter 7.3 and especially 7.3.4).

6.2   Tests and Results

6.2.1   UPILEX®

6.2.1.1   Mechanical Tests

To investigate the material behaviour under heat and vibration loads, tests using a dynamic mechanical analysis (DMA) were carried out with UPILEX® variants RN and SN [63]. A number of samples of each subtype was analysed at 1 Hz and a slowly increasing temperature to find the glass transition point (the temperature at which the material stiffness drops sharply) and the tolerable number of load cycles. Measurements of the storage and loss modulus are presented in Fig. 6.3. The second loss modulus peak, which coincides with a sharp drop in the storage modulus (the material's ability to return to its previous shape after the mechanical load is removed), marks the glass transition temperature. With ballute skin temperatures expected to reach as much as 200°C during the mission, both materials are suitable. The glass transition temperature of RN lies only marginally above the maximum expected mission temperature though, whereas SN can comfortably tolerate expected mission temperatures by a large margin.

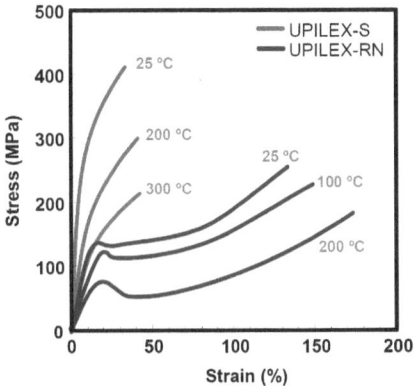

Fig. 6.3: Stress-Strain curves of UPILEX-S and -RN for selected temperatures [62]. (UBE)

Static stress strain tests were performed by simple pulling and under biaxial load at various temperatures [64]. The biaxial load was created by fixing a skin segment to a circular cavity with a clamp ring, pressurizing the enclosed volume with air. The test was done at room temperature and in an oven. Results match manufacturer specifications given in Fig. 6.3 of UPILEX [62] remarkably well and are in good agreement with tests performed at ESA ESTEC [65][66].

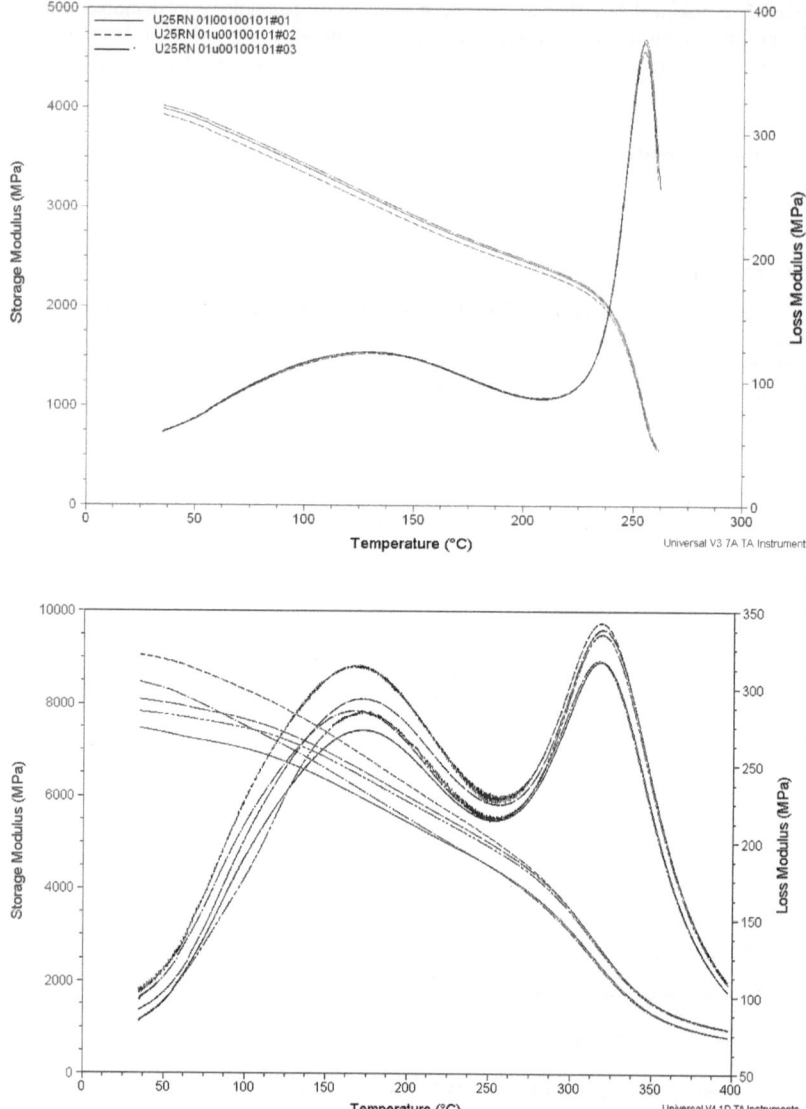

Fig. 6.4: Storage and loss modulus of UPILEX RN (top) and SN (bottom). [63]

6.2.1.2 Outgassing Tests

To validate the conformance of both UPILEX types to space qualification standards set forth in paragraph 7.2.3 of ECSS-Q-70-02A [56], Micro VCM tests were done at the DLR centre Berlin, Institute of Space Systems, on behalf of the Institute of Space Technology at the UBW and compared to earlier tests done by ESA ESTEC and stored in the ESMAT database [67]. The newer tests were necessary because the manufacturer of UPILEX, UBE, reported a change in the manufacturing process that might have had an effect on outgassing properties. No such problems were found, however.

Test results for UPILEX-S are TML=0.76%, RML=0.07% and CVCM=0%. Test results for UPILEX-RN differ only in RML, which is 0.1%. Currently available UPILEX variants therefore conform to space qualification requirements.

Outgassing tests with UPILEX-based adhesive tape manufactured by Lohmann Tapes were also performed by the DLR Berlin and showed that the silicone based adhesive layer fails to comply with requirements. Consultation with experts at the IABG space test centre in Ottobrunn, Germany, concluded that the total amount of adhesive used is too little to expect problems and that the amount directly exposed to vacuum (on either edge of the seam) is even less. Therefore the MIRIAM ballute was built using this Lohmann supplied high strength adhesive tape and later tested in the space simulation chamber at IABG. Routine measurements of the chamber's surface cleanliness revealed no deposits of any kind, leading to the conclusion that if the adhesive layer outgassed at all, deposits were below the measurement tolerance.

6.2.1.3 Thermo-Optical Properties

Since the thermal design of any spacecraft is based mainly on radiative heat exchange with its surroundings, knowledge of the thermo-optical properties is of paramount importance. Unfortunately, no such data exists yet for pristine, uncoated UPILEX [68]. Values for a range of metal-coated polyimides including UPILEX are presented by Fukuzawa et. al. [69].

Basic thermo-optical properties of untreated and uncoated UPILEX were therefore obtained with the help of ASTRIUM Ottobrunn, which allowed an employee to do the test free of charge, after regular working hours. A System 2000 model Spectrum GX FT-IR instrument was used in the test. Measurements were made for near infra-red, from 670nm wavelength to 3600nm. Results are presented in Fig. 6.6. Incidence angles were not varied during this test and neither was far-infra-red tested.

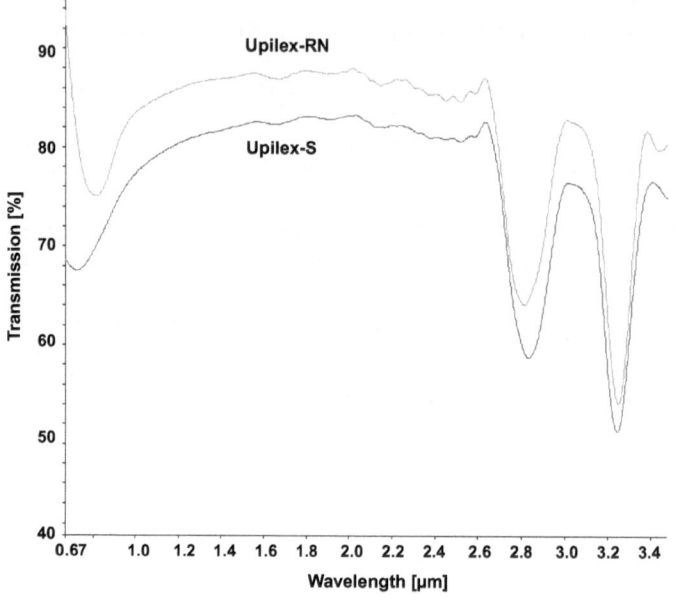

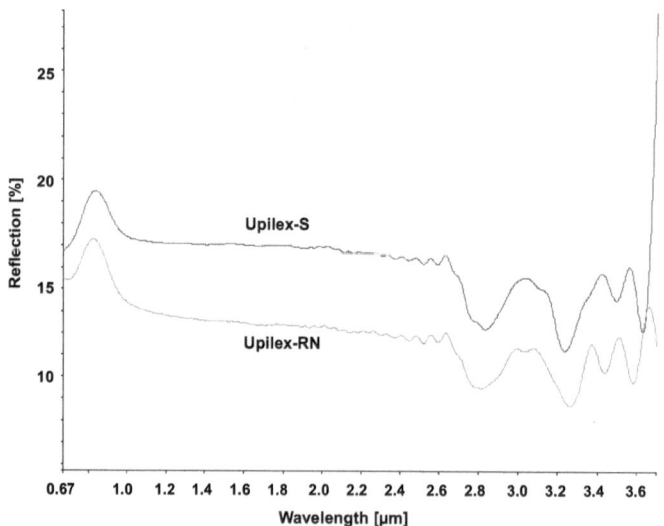

Fig. 6.5: Transmission (top) and reflection (bottom) coefficients for UPILEX from visible to near infra-red (Astrium).

6.2.2 PBO

In terms of mechanical and thermal resistance, Polyphenylbenzobisoxazole (PBO) outperforms any other plastic on the market today by far. It was developed in the 1980s by SRI International of Menlo Park, California, and later Dow Chemical, in an attempt to produce the strongest possible polymer chain by tying as many of the rugged benzene rings together as possible. The resulting polymer offered impressive performance, but exhibits long term stability issues that are until now not fully understood. On the commercial market it is currently only available as a fibre offered by Toyobo Co. Ltd. of Japan under the trade name "Zylon®" [70]. Unfortunately, it can only be obtained in form of either individual filaments or a spun yarn, although experiments with PBO thin film were carried out in the past. Foster-Miller of Waltham, Massachusetts has produced PBO thin film for experimental purposes and even manufactured a test balloon from it under a NASA contract. But the samples are comparatively small in size. They can presently be manufactured only as a tube of 3 in. diameter by 36 in. length and the process is more or less manual.

A major drawback of PBO is that it reportedly degrades under prolonged exposure to ultra-violet light and looses its mechanical properties under the influence of moisture [71]. A legal dispute over the application of PBO in body armour erupted after bullet proof vests made from Zylon® failed in two gun fights, mortally wounding one police officer and seriously injuring another [72][73]. These failures were attributed to the degradation of PBO under the aforementioned influences, halting wider spread use and mass production of PBO.

To investigate the underlying process and the validity of previous findings for ballute applications, tests at the UBW were carried out with Foster-Miller supplied PBO film, using Stress-Strain tests, DMA and FT-IR (Fourier-Transformed Infra-red Spectroscopy) on "virgin" and "degraded" samples [74][75][76]. Samples were subjected to uninterrupted storage for 2 years in a vacuum chamber (the thermal-vacuum chamber at the Institute of Space Technology at the UBW was used for this purpose). Others were stored in a temperature and humidity controlled container for various lengths in time and exposed to UV lamps.

Item	Unit	Regular	High Mod.
Sub-Type		AS	HM
Filament decitex	dtex	1.7	1.7
Density	g/cm3	1.54	1.56
Moisture Regain (65% RH)	%	2	0.6
Tensile Strength	GPa	5.8	5.8
Tensile Modulus	GPa	180	270
Elongation at Break	%	3.5	2.5
Melting Temperature		none	none
Decomposition Temperature in Air		650	650

Table 6.1: Material properties of Zylon® PBO fibres [70].

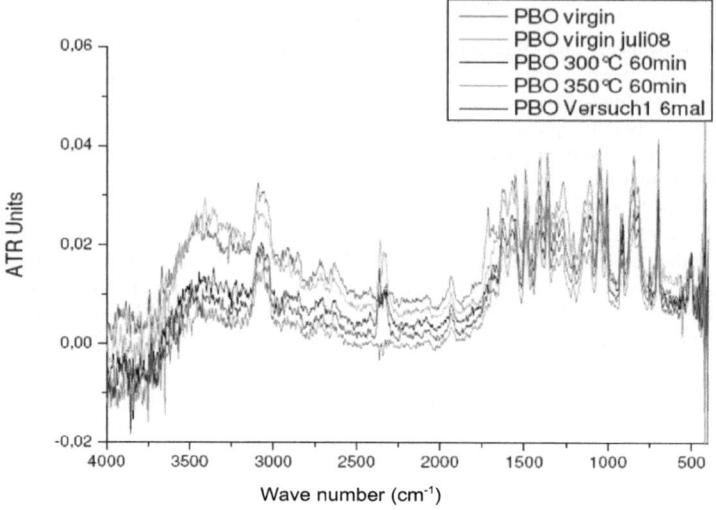

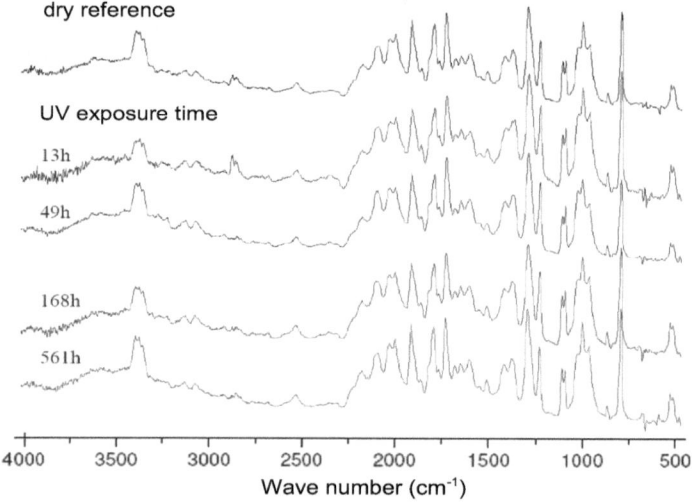

Fig. 6.6: FT-IR results of PBO after heat treatment (top) and UV exposure (bottom) [76].

For virgin and dry material, values presented in Table 6.1 could more or less be confirmed even for PBO thin film in stress-strain tests and are therefore of no greater concern.

FT-IR plots were obtained with the ATR-FTIR method (Attenuated Total Reflectance FT-IR) and show the chemical composition through a spectral analysis. Results (see sample analyses in Fig. 6.6) show no signs that either degradation method had any effect on the chemical composition of the PBO polymer, hinting that effects are not related to molecular decomposition at all and might not be permanent.

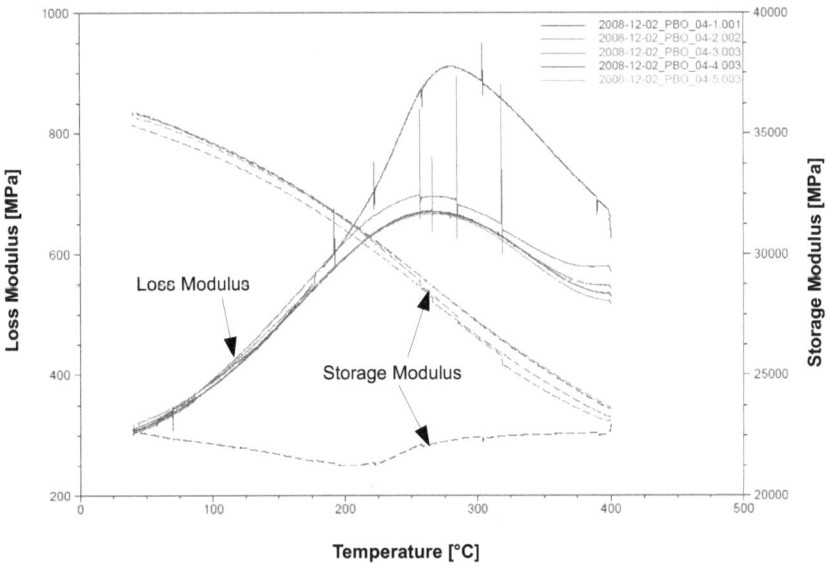

Fig. 6.7: Storage and loss modulus plots of several DMA passes with a single virgin PBO sample.

The degradation effect through moisture was indeed found to be mostly reversible. Samples subjected to a humid environment for an extended period of time showed a significant loss in mechanical strength, but regained almost all of it after it was "baked dry" in an oven. Fig. 6.7 shows the loss and storage modulus of a single PBO sample, obtained through repeated passes of a heated DMA cycle. It is interesting to observe that the material does not live up to expectations during the initial pass, but improves significantly in the second pass. Mechanical properties continue to improve throughout subsequent passes, which is due to loss of moisture during the heating process. This finding is of particular importance to ballute applications, because the ballute manufacturing and handling process is substantially easier if it may be subjected to a breathable atmosphere for most of the ground handling time. Before launching the package, it can be heated or other wise baked dry during the cruise phase of the mission.

In contrast to Toyobo's own findings as well as data obtained by NASA [71], results did not confirm a significant degradation as a result of exposure to UV and visible light [76]. While storage and loss modulus dropped about 19% within the first 200 hours of exposure to UV (see Fig. 6.8), values remain far above UPILEX limitations. The large difference to results of reference [71] may be explained by sample handling. Tests performed at NASA's Johnson

Space Center were made with samples that were stored at normal ambient conditions, drawing moisture from the humid air. Tests performed at the UBW, in contrast, were made with samples stored in a dry atmosphere controlled environment, to isolate the UV exposure effect from the moisture degradation effect. We may therefore assume that most of the UV-degradation found by [71] is indeed mostly the reversible moisture degradation effect. Note that the UV effect was not found to be reversible and must therefore be minimized during the manufacture by processing ballute parts in an environment with little UV radiation (such as in an ordinary clean room).

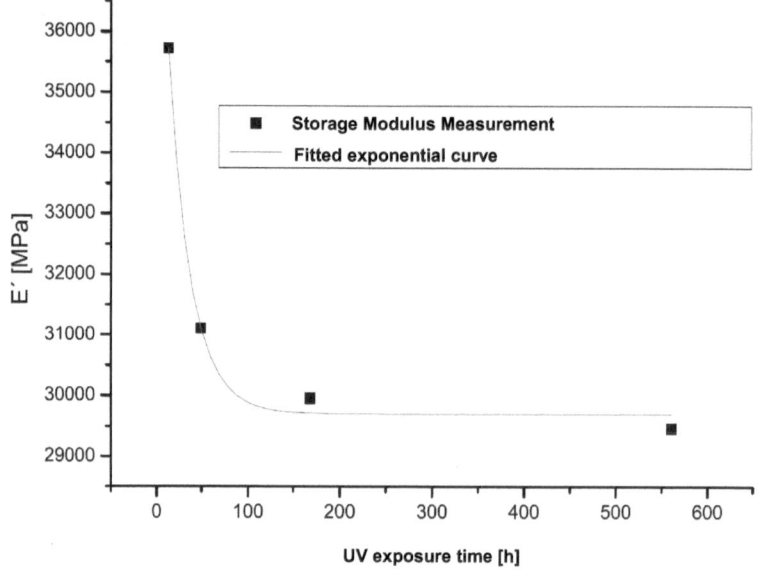

Fig. 6.8: Storage modulus of PBO film with respect to exposure time to UV radiation [76].

It should also be considered that a ballute spacecraft resides within its dark, temperature and atmosphere controlled storage container for the entire travel to its deployment point. Once deployed, it has a life expectancy on the order of tens of hours at most, whereas UV degradation happens on the time scale of hundreds of hours and humidity should not be a problem. The significance of UV degradation for the envisaged Mars mission is further reduced by considering that even the degraded material was much stronger during tests than any other material.

Packaging tests were not done with PBO, due to lack of sufficient material to build a test ballute. Deployment tests were not done either, except for pleat, folding and wrinkle tests with 7μm thick PBO film strips. Pleating and unfolding these strips showed no problem. However, film thickness has a direct relation to the sensitivity of a film to packaging damage, due to the higher geometrical moment of inertia. Thicker PBO film might therefore be expected to exhibit other properties and should be investigated for packaging damage, if it becomes a serious option.

No outgassing tests were carried out with PBO and neither were thermo-optical investigations.

6.3 Conclusion

ARCHIMATTER results obtained until today show that even heavily degraded PBO still offers a significantly superior, almost metal like performance, as compared to any other polymer.

However, large amounts of PBO film, especially in large pieces, remain unobtainable.

Tests with UPILEX showed that UPILEX-S is the best alternative in terms of mechanical strength, but that UPILEX-R is the only version of the polyimide that can be packed and deployed without damaging the ballute envelope.

Because UPILEX-R alone does not offer sufficient margin to fly the mission, a ballute envelope exclusively made of UPILEX-R will have to be reinforced. Such reinforcement can be provided by the seams in a beach ball pattern, which channel forces and can be made of a slender PBO band (see chapter 7.3.4.3).

If a PBO ballute is considered, but large sheets of PBO cannot be extruded, a geodesic sphere production pattern may be chosen, at the expense of production complexity and more difficult seeming (see chapter 7.3.2.2).

7 Ballute

The ballute's purpose is to set the size and aerodynamic properties of the spacecraft, so that its entry trajectory and attitude along the flight path conform to mission requirements.

7.1 System Design

The ballute is an inflatable envelope, in our case a spherical thin film balloon. In addition to the main body, the ballute needs an inflation gas distribution system, an inflation gas duct and an interface to the service spacecraft,. which provides a releasable mechanical connection and gas throughput.

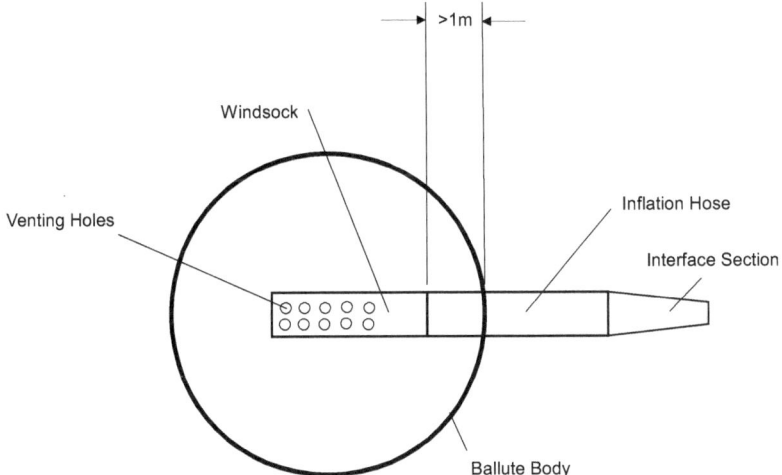

Fig. 7.1: General layout of the spherical ballute showing all functional elements (except the Instrument Pod).

For the gas distribution system, we choose the standard method of a windsock. Windsocks are normally fabric bags that vent through the fabric pores. In our case, the windsock is a perforated thin-film bag that is closed at the end and has a multitude of venting holes (see Fig. 7.1). This prevents vibrations and buffeting through boundary layer separation at the end of the inflation hose. A constant windsock overpressure keeps it stiff and and the gas flow steady. Through the little venting holes, the inflation gas is uniformly and evenly distributed inside the ballute.

For a gas duct we also choose a standard balloon method, namely a thin film inflation hose that passes through a slit in the ballute skin (see Chapter 9). Under overpressure, the inflation hose will be stiff and keep the slit open. If gas supply pressure drops, inflation hose and windsock will collapse and automatically seal the ballute. This is standard balloon practice and needs no further theoretical treatment.

The general layout of the spherical ballute, showing all relevant elements, is given in Fig. 7.1. The 36-segment 10-m ARCHIMEDES ballute is also presented in Fig. 7.47 on page 131 and the MIRIAM flight model ballute is shown in Fig. 10.5 on page 188.

The ideal spherical ballute would be a perfectly round thin film envelope, extruded like a chewing gum bubble. While theoretically possible, this is practically infeasible, as no plant or factory exists that is geared towards extruding and processing large thin film bubbles. So the practical ballute design has to work with rolls of thin-film of available shapes and sizes. We therefore seek to build something large and round out of something small and flat (see Chapter 7.3.2).

Since any real world ballute with all its seams, attachment points, windsock, inflation hose and possibly even instrumentation will be inevitably heavier than a perfect thin film bubble of equivalent dimensions, we can define a "production mass" m_{pf} that accounts for all masses which do not belong to the perfect skin sphere. We can then set the two masses into relation to define a production mass factor or in short production factor:

$$\frac{m_s}{m_{pf}} = \lambda_{pf} \tag{56}$$

For further treatment of this factor refer to Chapter 7.2.4.

7.2   Theory of Operation (Ballute Theory)

The ballute decreases the ballistic coefficient of the spacecraft by increasing the drag effective area and the hypersonic drag coefficient, through its shape and its overall low mass density. The shape also influences the location of the spacecraft's centre of mass, its moments of inertia and its aerodynamic lift and momentum coefficients.

The ballute theory developed for this analysis is genuine, although several very good publications address the theoretical treatment of the matter in many different ways (such as [77] and [36], to name just two). Unfortunately, no publication by the Goodyear Corporation on the theoretical treatment of ballute design could be located. It is, however, believed that a similar theory must exist, if perhaps beyond the reach of readily accessible catalogues.

In this chapter, we will explore the principal theory underlying the design of a ballute that is tailored to mission requirements and bound by real world constraints. We start with the determination of the ballistic coefficient by using a simplified analytical model, which is rather imprecise, but still good and simple enough for use in a pre-phase A study. We also discuss model adaptability issues for high altitudes and shallow entry angles, two key characteristics. We will then proceed with the inflation pressure requirements and constraints and continue with an equally approximative descent analysis. Finally, we will address ballute sizing and design considerations.

As a general simplification of the problem, we may assume that the ballute is spherical and produces no aerodynamic lift. For ARCHIMEDES and the flight test MIRIAM (see chapters 10.2 and 10.3), this is an exact match. But even for a Goodyear-derived shape, the deviation from a perfect sphere is small enough to work reasonably well in a first-order-of-magnitude parametric analysis. In case a conical shape such as IRDT or a toroidal or lentil shape is chosen, the mathematical equations must be derived anew, possibly even including lift, but the example given herein may still serve as a procedural guideline.

For a spherical ballute spacecraft, we can define the following basic relations.

Hull area:

$$A_H = 4\pi r_b^2 \tag{57}$$

Spacecraft drag effective face area:

$$A_{SC} = \pi r_b^2 \tag{58}$$

Spacecraft (ballute) volume:

$$V_{BL} = \frac{4}{3}\pi r_b^3 \tag{59}$$

In (59) we assume that the volume of the instrument pod has little to contribute to the overall spacecraft volume. Because for any useful ballute mission with a sufficiently low ballistic coefficient, $V_{ip} \ll V_{BL}$ should always be true.

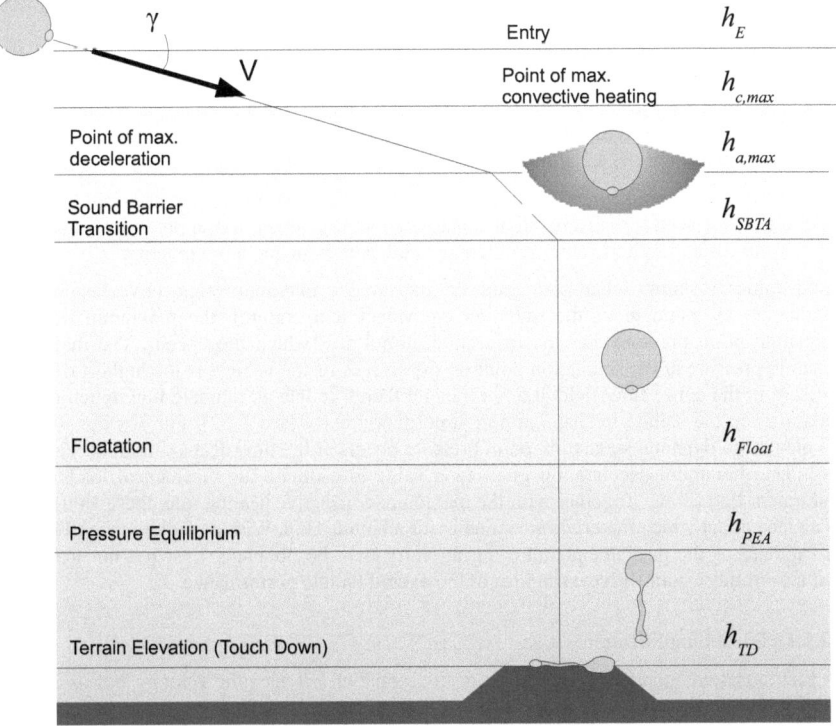

Fig. 7.2: Definition of characteristic events and altitudes during a ballute mission without lift.

As far as mission design is concerned, the most interesting reference points for a simple parametric analysis are deemed to be the following (see Fig. 7.26):

1. Entry into the atmosphere (denoted E);

2. Maximum Convective Heating Rate (denoted c,max);

3. Maximum Deceleration (denoted a,max);

4. Sound Barrier Transition (denoted SBTA;)

5. Floatation (denoted Float);

6. Pressure Equilibrium Altitude (denoted PEA);

7. Touch Down (denoted TD).

During an actual mission analysis process, one may see the need to define an additional altitude of choice. This is well possible but will not be treated separately in this chapter. Of course this chapter can indeed serve as a procedural guideline.

7.2.1   An Approximate Analytical Method to Determine the Desired Ballistic Coefficient and Ballute Performance

Ideally a thorough trajectory and mission profile analysis should precede the outset of the ballute design phase, so that all requirements, including the desired ballistic coefficient and Instrument Pod mass are known within a reasonable range. Such an analysis should include studies of numerically integrated or analytically determined trajectories of some degree, as put forward in chapters 3, 4.3 and 5.

In an iterative process, however, it might be of interest to first investigate principal technical capabilities. As such, precision might not be of much concern and approximate analytical solutions to the atmospheric entry problem are both sufficient and more practical.

Good values to know when sizing the ballute are the maximum convective heating rate (leading to an estimate of the radiation equivalent temperature), the maximum dynamic stagnation point pressure and the respective altitudes at which they occur. The maximum dynamic pressure at the stagnation point on the surface of the vehicle is the highest dynamic pressure in the entire flow field at any time and therefore lets us estimate how much gas we need to keep the ballute in shape during atmospheric entry (see 7.2.2). For obvious reasons, the maximum dynamic stagnation point pressure occurs at the time of maximum deceleration [36]. The maximum deceleration gives us a value to estimate the mechanical loads at the Instrument Pod fitting. Together with the maximum convective heating rate, these values give us an idea about which materials are suitable for a ballute skin. With its radiation coefficients, an estimate of the peak temperature during entry may be obtained. Last but not least, the altitudes of these points give us an idea of the overall ballute performance.

7.2.1.1 Basic Simplifications

To find the above mentioned values, we desire a relation between the velocity and its change over time and altitude. Let's assume for this purpose that the ballute spacecraft enters a straight and planar gas field with an isothermal exponential density gradient such as the one presented in chapter 3.3. Inside the gas field, it encounters a dynamic pressure q_∞ given by [29]:

$$q_\infty = \frac{1}{2} \rho_\infty v_\infty^2 \qquad (60)$$

where ρ_∞ is the local ambient density, not influenced by the passing spacecraft, and v_∞ the undisturbed gas flow velocity. If the gas field is assumed to be stationary, v_∞ is also the vehicle velocity. The local ambient density is obtained from the exponential density relation of an isothermal atmosphere model:

$$\rho_\infty = \rho_0 \cdot e^{\frac{-h}{H_0}} \qquad (61)$$

where ρ_0 the atmospheric density at the "bottom" of the gas field, H_0 the gas field's characteristic density scale height and h the altitude above the bottom (see chapter 3.3 for more details).

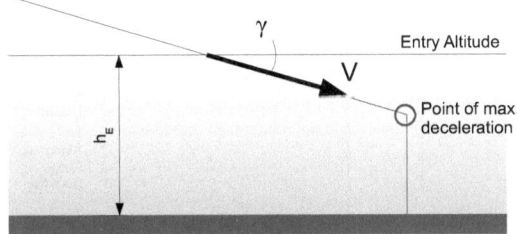

Fig. 7.3: Simplification of the Entry Problem.

We further assume that the spacecraft flies through the gas without markedly changing its flight path angle [27], until it reaches the point of maximum deceleration (see Fig. 7.3 and refer to chapter 3.2 for a definition of the flight path angle). This is actually accurate for the case where the flight path angle is -90° and no dynamic lift is produced and a reasonable simplification for all flight path angles of around -15° and steeper . The reason for this is that the velocity from orbital dynamics is so high that gravitational and centrifugal forces balance each other out sufficiently for the object to appear flying straight [27] until the aerodynamically induced $\frac{dv}{dt}$ is so large that centrifugal forces become too weak to counteract gravity.

For an approximate analytical description of the trajectory up to the point of maximum deceleration, the gravity force can therefore be eliminated from the basic equations of motion as defined in chapter 3.2 [27]. Since the following change in flight path angle happens quite rapidly, the segment of the trajectory which reaches from this point to the ground is almost straight down. This latter fact bears much resemblance to an egg which is hurled at a wall, perhaps in anger, but anyway fast enough to appear flying more or less straight until it "smacks" into the wall. There it suffers a short but large $\frac{dv}{dt}$ and then just drops to the floor.

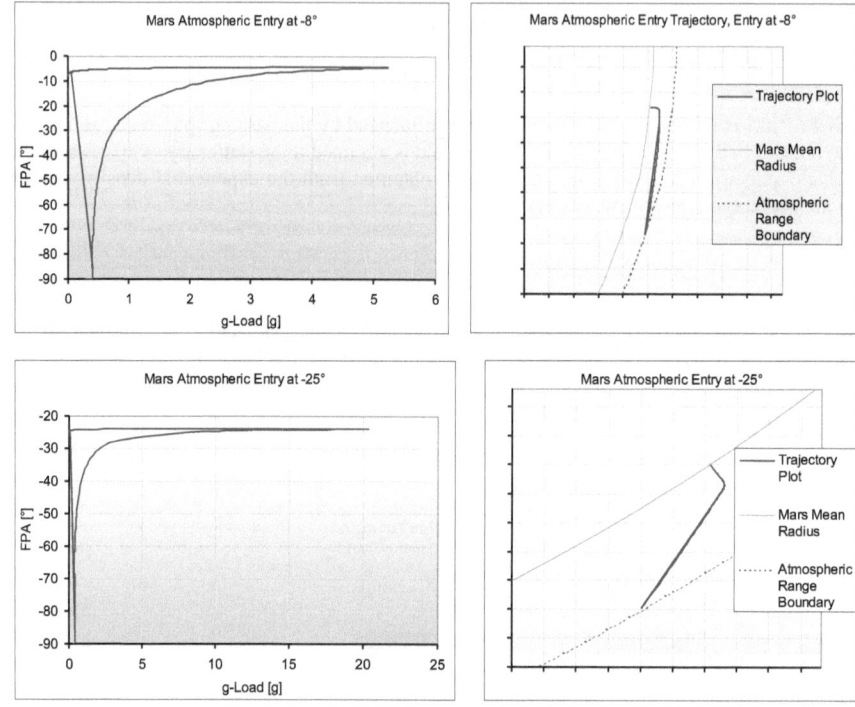

Fig. 7.4: Numerical computation of the flight path angle versus deceleration (in Earth g's) for two example entries at -8° and -25°. (left: 2D trajectory plots).

To exemplify the case and cross check with numerical analysis, two trajectories of a vehicle with a ballistic coefficient of around 0.5 kg/m² and entry angles of -8° and -25° were computed with the code based on the RSS. The results are depicted in Fig. 7.4. Clearly, the flight path angle does not depart more than a few degrees from the entry angle until the point of maximum deceleration. On the 2D plots in Fig. 7.4, the point of maximum deceleration really looks like the vehicle "collides" with something and then falls almost straight down.

An important limitation of this method is that skipping entries cannot be treated this way, as the gas field is assumed straight and planar, instead of being curved. Therefore even very shallow angles lead to a direct entry result, which is of course not realistic. Entry angles more shallow than a certain limit, $\gamma_{shallow,\,limit}$ are not recommended for investigation by this method. The actual value of $\gamma_{shallow,\,limit}$ depends on the gravitational acceleration of the planet, its atmosphere parameters and the vehicle's entry speed. In case of a typical Mars entry, the limit is around -7°.

The interested reader might want to note that a similar simplification of the entry problem, along with a host of other models, is also treated by [27].

7.2.1.2 The Velocity as a Function of Altitude

We start by seeking a basic relation between the velocity of the spacecraft as a function of its instantaneous altitude. From our basic equations of motion and particularly equation (18) we know that the velocity change is

$$\frac{dv_{sc}}{dt} = -\frac{1}{2}\frac{\rho_\infty}{\beta} \cdot v_{sc}^2 \tag{62}$$

and from the simplification at hand follows that the altitude change is

$$\frac{dh}{dt} = -v_{sc}\sin\gamma \tag{63}$$

with the flight path angle $\gamma \approx \gamma_E = const.$. As stated before the gravity force can be neglected here in a first order of magnitude estimate, because centrifugal and gravity forces mostly cancel each other out [27]. Since the spherical ballute is symmetric and smooth, aerodynamic lift can also be neglected. We can now combine equations (61), (62) and (63) into the following relation between velocity and altitude:

$$v\frac{dv}{dh}\sin\gamma = \frac{\rho_0 e^{-h/H_0}}{2\beta}v^2 \tag{64}$$

This equation can be rearranged so that

$$\int_{v_E}^{v_{sc}}\frac{dv}{v} = \int_{h_E}^{h_{sc}}\frac{\rho_0 e^{-h/H_0}}{2\beta\sin\gamma}\,dh \tag{65}$$

The right side of (65) now represents the density gradient that the spacecraft encounters along its trajectory, whereas the left side represents the velocity change. Note that the right side already represents a gradient in the relation between the spacecraft's density and that of the surrounding atmosphere, as pointed out in chapter 3.4. This easily integrates into

$$\ln v_{sc} - \ln v_E = \frac{H_0\rho_0}{2\beta\sin\gamma}\left(e^{\frac{-h_E}{H_0}} - e^{\frac{-h_{sc}}{H_0}}\right) \tag{66}$$

Since we allowed the drag coefficient (and therefore the ballistic coefficient) and the flight path angle to be constant until the point of maximum deceleration is reached, we can further simplify treatment of the subject by combining all the "constants" into a single, dimensionless trajectory parameter ϑ :

$$\vartheta = \frac{H_0\rho_0}{2\beta\sin\gamma} \tag{67}$$

If you will, this trajectory parameter now finally represents the relation between the "effective" density of the spacecraft (β) and the atmosphere $(H_0\,\rho_0)$ and a factor for its rate of change $(\sin\gamma)$. The velocity of the spacecraft may therefore be written as a function of the instantaneous altitude of the spacecraft:

$$v_{sc} = v_E \, e^{\vartheta\left(e^{\frac{-h_g}{H_g}} - e^{\frac{-h_s}{H_s}}\right)} \tag{68}$$

This equation is the basic tool we need for almost all of the following analyses.

7.2.1.3 Maximum Deceleration

The first value we look for is the point of maximum deceleration, which we find by setting its time derivative to zero [27]. To do this, we combine (62) with (68) to obtain the space craft acceleration in the following form:

$$\frac{dv_{sc}}{dt} = a_{sc} = -\frac{\rho_0 e^{\frac{-h_{sc}}{H_0}}}{2\beta} v_E^2 \, e^{2\vartheta\left(e^{\frac{-h_g}{H_g}} - e^{\frac{-h_s}{H_s}}\right)} \tag{69}$$

The time derivative of the acceleration may then be found through

$$\begin{aligned}
\frac{a_{sc}}{dt} &= -\frac{\rho_0 v_E^2}{2\beta}\frac{d}{dh}\left\{ e^{2\vartheta\left(e^{\frac{-h_g}{H_g}} - e^{\frac{-h_s}{H_s}}\right)} - e^{\frac{-h_{sc}}{H_0}}\right\} \\[2mm]
&= -\frac{\rho_0 v_E^2}{2\beta H_0}\left\{ e^{\frac{h_{sc}}{H_0} + 2\vartheta\left(e^{\frac{-h_g}{H_g}} - e^{\frac{-h_s}{H_s}}\right)}\left(1 - 2\vartheta\, e^{\frac{-h_{sc}}{H_0}}\right)\right\}
\end{aligned} \tag{70}$$

Setting this equation zero yields a compact expression for the altitude at which maximum deceleration occurs:

$$\frac{da_{sc}}{dt} = 0 \;\Rightarrow$$

$$h_{a,max} = H_0\cdot\ln\left(2\vartheta\right) \tag{71}$$

$$= H_0\cdot\ln\left(\frac{H_0\rho_0}{\beta\sin\gamma}\right)$$

If mission design wishes to choose the approximate maximum deceleration altitude, we can solve (71) for a "design ballistic coefficient":

$$\beta_{design} = \frac{H_0\rho_0}{\sin\gamma}\, e^{-\frac{h_{a,max,design}}{H_0}} \tag{72}$$

Note that (72) does not say anything about whether the desired ballistic coefficient is technically achievable! This is discussed in in chapter 7.2.4.

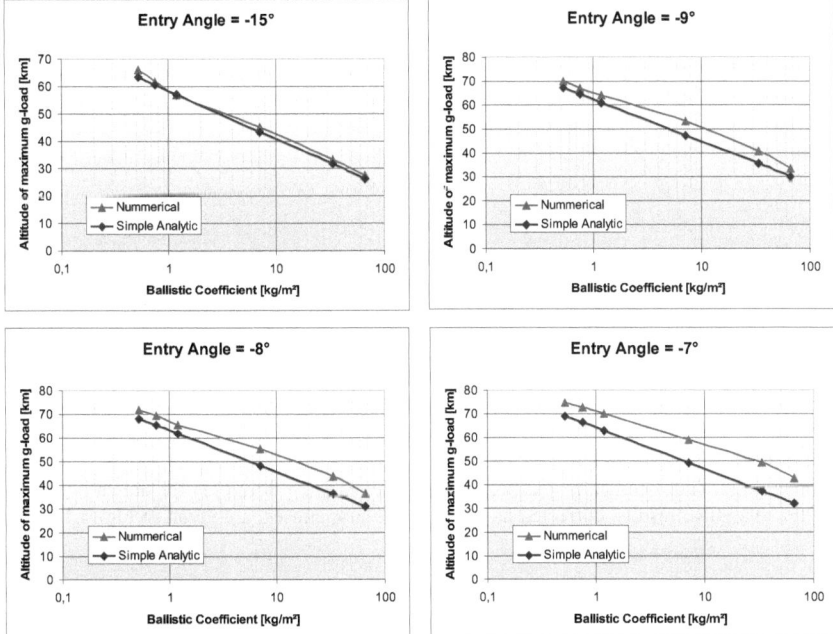

Fig. 7.5: The altitude of maximum deceleration for a Mars entry with respect to ballistic coefficient from a simple analytical solution with no correction for shallow entries as compared to numerical trajectory calculation results based on a daytime atmospheric profile of the Mars Climate Database [32].

Fig. 7.5 gives results obtained with the above model, as compared to results obtained with a numerical trajectory simulation based on the RSS and a Mars Climate Database profile [32]. To improve the barometric gas field model, a corrected barometric atmosphere model, matched to high altitudes as described in chapter 3.3, was used.

Still, it is obvious that the results are as imprecise as they come, especially for very shallow entry angles, where the approximation of a constant flight path angle is not a very good one. The reason is that the long trajectory segment arches through the sky above a curved horizon and is long enough to render the assumption of an invariant flight path angle invalid. In Fig. 7.4, we can already see that for an entry angle of -8° the flight path angle decreases to almost half the value of the original angle at entry.

We therefore seek to find an equivalent constant FPA, to extend the usable range of this model for the maximum deceleration altitude to shallow entry angles (for Mars this is generally between -7° and -15°). The following relation between the entry angle and the flight path angle was found to work well:

$$\gamma = \gamma_{mean, const} \approx \tanh\left[\left(\gamma_E - 2\right)^2\right] \cdot \gamma_E = const. \tag{73}$$

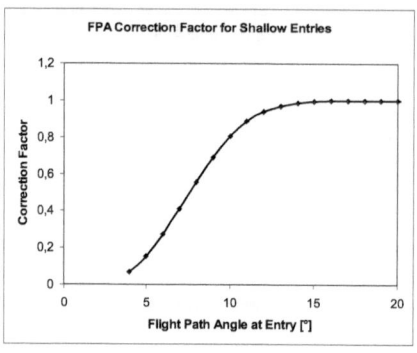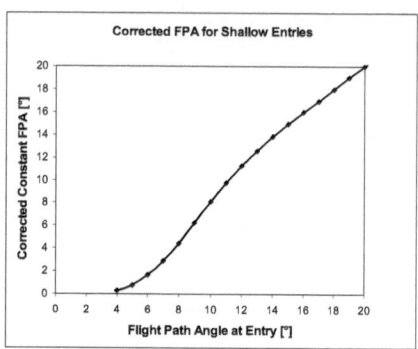

Fig. 7.6: Correction factor and corrected flight path angle for shallow entries plotted against the flight path angle at the atmospheric interface.

This was found by fitting the equivalent flightpath angle to a mean value of the real flight path angle plot.

The correction factor and corrected flight path angle resulting from (73) are shown in Fig. 7.6 and the analytical results, corrected as such for shallow entries, are given in Fig. 7.7. These

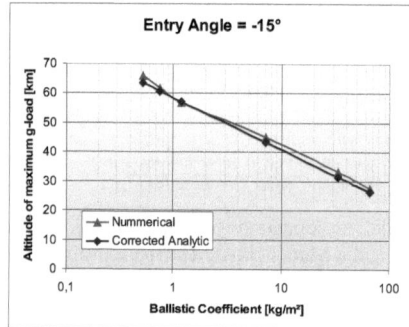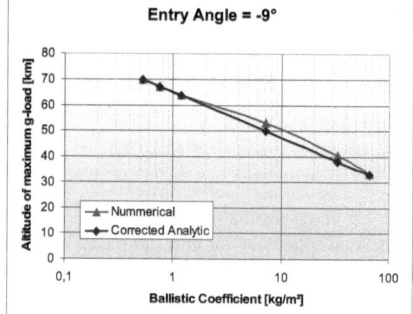

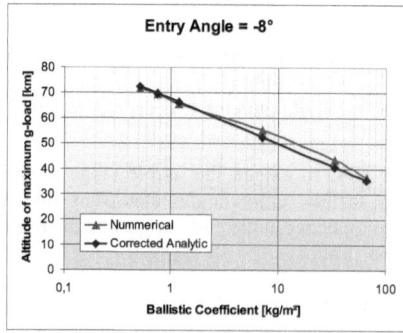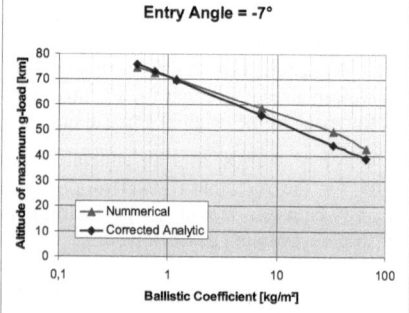

Fig. 7.7: The altitude of maximum deceleration for a Mars entry with respect to ballistic coefficient from a corrected analytical solution, as compared to numerical trajectory calculation results obtained with the RSS.

results are already better than many of the other expected errors, not the least of which are changes in atmospheric profiles due to landing coordinates, weather and seasonal influences, effects that can hardly be known at the outset of a pre-phase A study.

The maximum acceleration can now be found by combining (69) and (71) to obtain:

$$a_{sc,max} = \frac{v_E^2 \sin \gamma}{2H_0} \exp\left[-1+\left\{\frac{H_0 \rho_0}{\beta \sin \gamma} e^{-\frac{h_E}{H_0}}\right\}\right] \tag{74}$$

Under all expected mission conditions, the term $\dfrac{H_0 \rho_0}{\beta \sin \gamma} e^{-\frac{h_E}{H_0}}$ is very small, so we may further simplify:

$$a_{sc,max} \approx \frac{v_E^2 \sin \gamma}{2H_0 e} \tag{75}$$

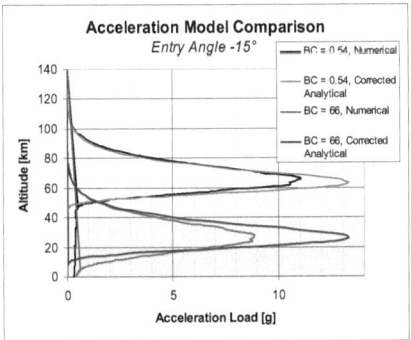
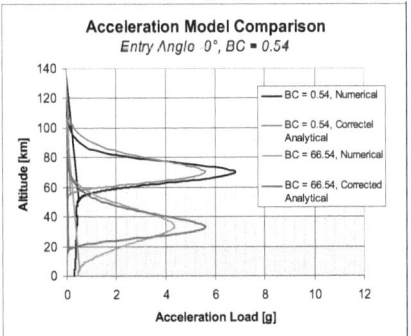

Fig. 7.8: Comparison of analytical and numerical acceleration curves upon entry into the Martian atmosphere for various ballistic coefficients. Numerical results are based on a daytime Mars Climate Database.

It is interesting to observe that in this simple model, the maximum deceleration value is almost independent from the ballistic coefficient (safe for a negligible contribution) and therefore depends almost solely on the entry speed and the entry angle! Cross-checking with numerical computations based on the RSS shows us that this is not really so, as can be seen in Fig. 7.8, Fig. 7.9 and Fig. 7.10. Care must therefore be taken when evaluating the above equation for maximum deceleration and a real design must be based on proper numerical trajectory computations, as the analytical model presented herein is clearly an oversimplification and not suitable to determine actual mission design loads. Fig. 7.8 also shows that the planet's own gravitational acceleration has been omitted in this model, a simplification which is not acceptable any more for trajectory points significantly after the point of maximum deceleration.

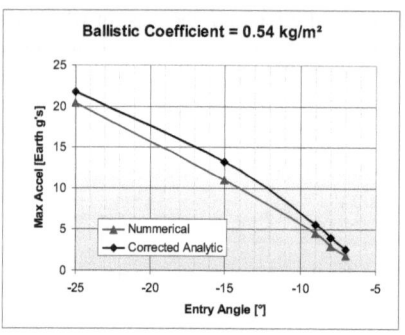 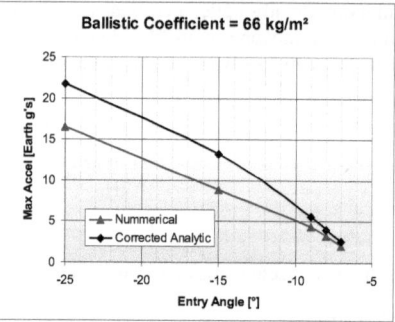

Fig. 7.9: Comparison of analytical and numerical maximum acceleration computations for an entry into the Martian atmosphere for two different ballistic coefficients.

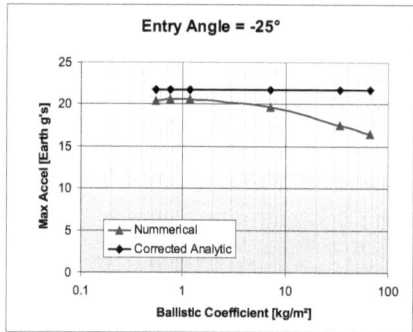 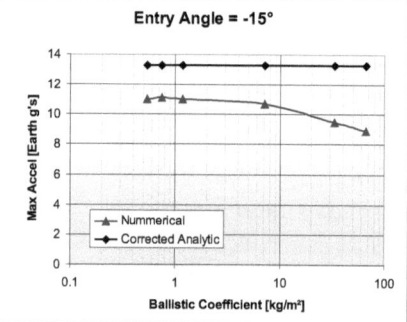

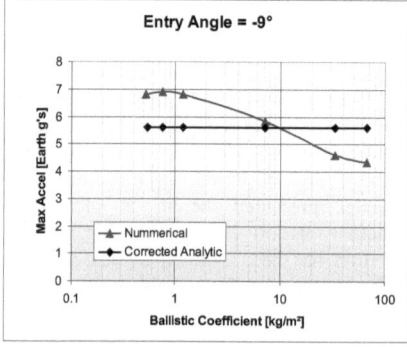 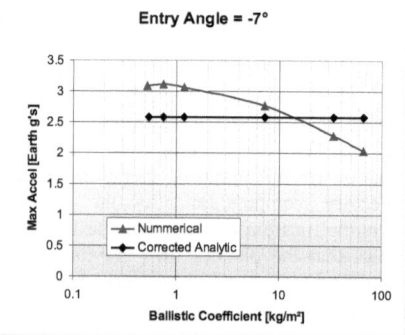

Fig. 7.10: Comparison of analytical and numerical maximum acceleration computations for an entry into the Martian atmosphere for four different entry angles .

7.2.1.4 Maximum Hypersonic Heating Rate

The highest heat influx driving a ballute development is the heat resulting from friction and the compression wave during hypersonic flight. To find the maximum convective heating rate we consult references [37] and [39]. This occurs at the stagnation point at the nose tip of the vehicle and was determined by Sutton & Graves [37] to be

$$\dot{q}_r = \frac{\tilde{k}}{10^5} \sqrt{\frac{\rho_\infty}{R_N}} \, v_{sc}^3 \quad \text{in} \quad \left[\frac{W}{m^2}\right] \tag{76}$$

with $\tilde{k} = 18.9$ as the gas mixture factor for most of the Martian and Venusian atmospheres and $\tilde{k} = 18.1$ for most of the Earth's [37]. We obtain ρ_∞ from our atmosphere model, then v_{sc} from equation (68) and neglect $e^{-\frac{h_E}{H_0}}$ because it is so small and arrive at the following expression:

$$\dot{q}_c = \frac{\tilde{k} \, v_E^3}{10^5} \sqrt{\frac{\rho_0}{R_N}} \exp\left[-\frac{1}{2}\frac{h_{sc}}{H_0} - 3.9\,e^{-\frac{h_{sc}}{H_0}}\right] \tag{77}$$

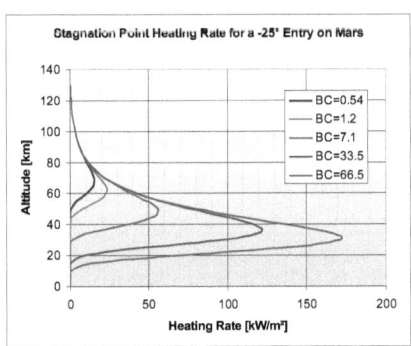

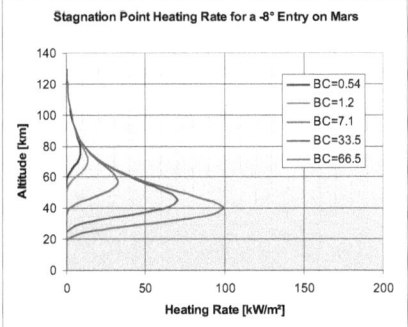

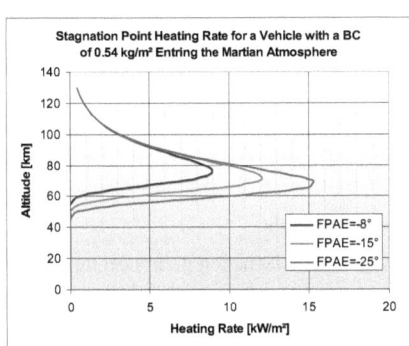

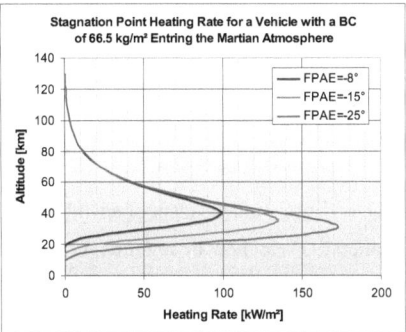

Fig. 7.11: Analytical calculation of the stagnation point heating rate over entry altitude for a vehicle entering the Martian daylight atmosphere at 4.5 km/s.

This equation is evaluated for a range of ballistic coefficients and entry angles and given in Fig. 7.11. Such is the nature of this equation that the altitude of the maximum heating during the mission depends a lot more on the ballistic coefficient than on the entry angle.

This fact becomes clearer perhaps when considering that the trajectory parameter ϑ has the ballistic coefficient as a direct contributor and the flight path angle as the argument of a sine function.

Note that these equations only give an estimate of the hypersonic heat input. A thermal analysis based on the thermo-optical properties of candidate materials may be done as part of the skin material selection process to determine how much energy is radiated back into space and ultimately which temperature the ballute surface will assume.

Finding the altitude at which the maximum heating rate occurs is analogous to finding the altitude of the maximum deceleration. With equation (77) we may again build a time derivative:

$$
\begin{aligned}
\frac{\dot{q}_c}{dt} &= \frac{\tilde{k}\, v_E^3}{10^5} \sqrt{\frac{\rho_0}{R_N}} \frac{d}{dh} \exp\left[-\frac{1}{2}\frac{h_{sc}}{H_0} - 3\vartheta\, e^{-\frac{h_{sc}}{H_0}}\right] \\
&= \frac{\tilde{k}\, v_E^3}{10^5} \sqrt{\frac{\rho_0}{R_N}} \left(e^{\frac{h_{sc}}{H_0}} - 6\vartheta\right) \exp\left[-\frac{3}{2}\frac{h_{sc}}{H_0} - 3\vartheta\, e^{-\frac{h_{sc}}{H_0}}\right]
\end{aligned}
\tag{78}
$$

Setting this zero yields the following handy equation for the altitude of maximum stagnation point heating:

$$
\frac{\dot{q}_c}{dt} = 0 \;\Rightarrow
$$

$$
h_{c,max} = H_0 \cdot \ln(6\vartheta) = H_0 \cdot \ln\left(\frac{3 H_0 \rho_0}{\beta \sin\gamma}\right)
\tag{79}
$$

Interestingly, from equation (71) and (79) we can follow that the maximum heating rate for a non-skipping entry lies always above the point of maximum deceleration and that the difference between the two is always

$$
h_{c,max} - h_{a,max} = H_0 \ln(6\vartheta) - H_0 \ln(2\vartheta) = H_0 \ln(3)
\tag{80}
$$

which is entirely independent of anything except the scale height.

Equation (77) can now be rewritten to give us the maximum stagnation point heating rate, by inserting (79) in place of the spacecraft altitude:

$$
\dot{q}_{c,max} = \frac{\tilde{k}\, v_E^3}{10^5} \sqrt{\frac{\rho_0}{6 R_N \vartheta e}} = \frac{\tilde{k}\, v_E^3}{10^5} \sqrt{\frac{\beta \sin(\gamma)}{3 e H_0 R_N}}
\tag{81}
$$

Equations (79) and (80) are compared against numerical trajectory integration results based on the Mars Climate Database in Fig. 7.12.

According to this evaluation the altitude of the peak value is in good agreement with numerically obtained values. Care must anyway be taken when these values are applied to sizing the ballute skin, as the stagnation point heating model by Sutton & Graves itself is only an approximation [37]. No real flight data for vehicles with such low ballistic coefficients is currently available.

Nonetheless, we can already see that a lower ballistic coefficient also leads to a lower amount of heat which is transferred to the spacecraft. This is one of nature's more favourable treats, because the lighter a possible skin film gets, the lower its tolerable heating rate is.

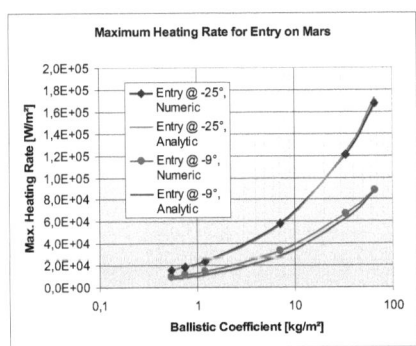
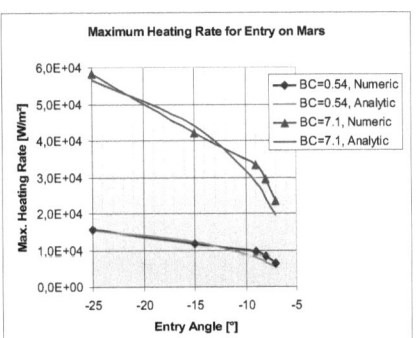

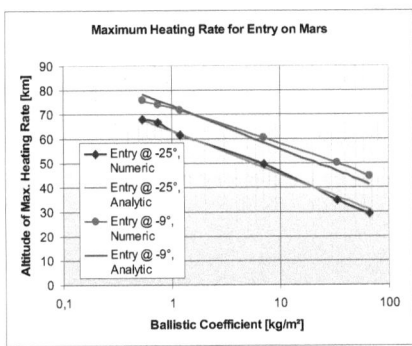
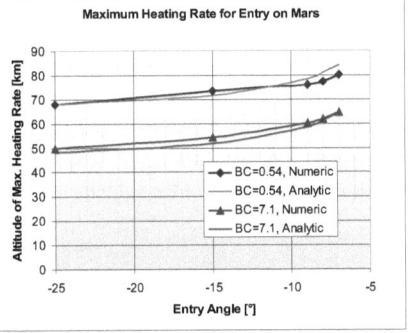

Fig. 7.12: The stagnation point convective heating model by K.Sutton and R.Graves [37], used in a numerical code based on the Mars Climate Database [32], as compared to its use in a simple analytical model based on a matched isothermal model.

We can use equation (81) to specify the ballistic coefficient we need to achieve if we know the maximum tolerable heating rate of a given material (from the maximum tolerable temperature and the thermal radiation properties).

$$\beta_{design,\,max} = \dot{q}_{c,\,max}^2 \cdot \frac{3e\,10^{10}\,H_0\,R_N}{v_E^6\,\tilde{k}^2\sin(\gamma)} \tag{82}$$

This equation is plotted in Fig. 7.13 for a 10-metre ARCHIMEDES-type ballute entering at different flight path angles. Evidently, aspiring the longest possible descent time on Mars (requiring the lowest possible ballistic coefficient) still puts modern thin film technology to the ultimate test. One might feel tempted now to argue that if a low ballistic coefficient is required anyway, a low temperature resistant thin film material may be chosen as well. Unfortunately, however, lighter materials also have less mechanical strength and their mechanical properties degrade faster with an increasing heating rate, which might render them unsuitable to withstand mechanical loads.

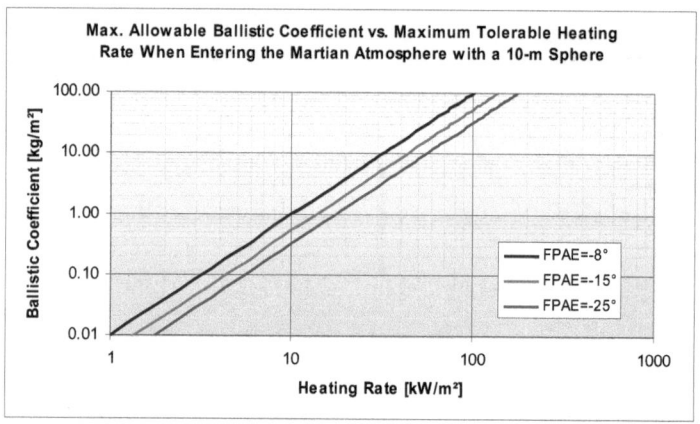

Fig. 7.13: Maximum allowable ballistic coefficient vs. the maximum tolerable heating rate of a given ballute skin material.

7.2.1.5 Maximum Stagnation Point Pressure

The maximum stagnation point pressure can be found by recomposing the dynamic pressure equation (60) in terms of the velocity at which maximum deceleration occurs:

$$q_{max} = \frac{1}{2} \rho_{\infty, a, max} v_{a, max}^2 \tag{83}$$

Note that in Equation (83) q represents the dynamic pressure and must not be confused with the convective heating rate. Equation (68) gives the velocity at the point of maximum deceleration:

$$v_{a, max} = v_E \exp\left[\vartheta \left(e^{\frac{-h_E}{H_0}} - e^{\frac{-h_{a, max}}{H_0}} \right) \right] \tag{84}$$

With the maximum deceleration altitude from equation (71) this becomes

$$v_{a, max} = v_E \exp\left[-\frac{1}{2} + \vartheta\, e^{\frac{-h_E}{H_0}} \right] \tag{85}$$

Again, the term $\vartheta e^{-\frac{h_E}{H_0}}$ is very small under all expected mission conditions, so we may further simplify:

$$v_{a,max} = v_E e^{-\frac{1}{2}} = \frac{v_E}{\sqrt{e}} = 0.6065 \cdot v_E \tag{86}$$

With (85) the stagnation point pressure becomes

$$q_{max} = \frac{1}{2}\rho_0 e^{\frac{-h_{a,max}}{H_0}} v_E^2 \exp\left[-1 + 2\vartheta e^{\frac{-h_E}{H_0}}\right] \tag{87}$$

which we may further simplify to:

$$q_{max} \approx \frac{\rho_0 v_E^2}{2} e^{-\frac{h_{a,max}}{H_0} - 1} \tag{88}$$

Results of this dynamic pressure relation are compared against numerical computation results based on the Mars Climate Database [32] in Fig. 7.14.

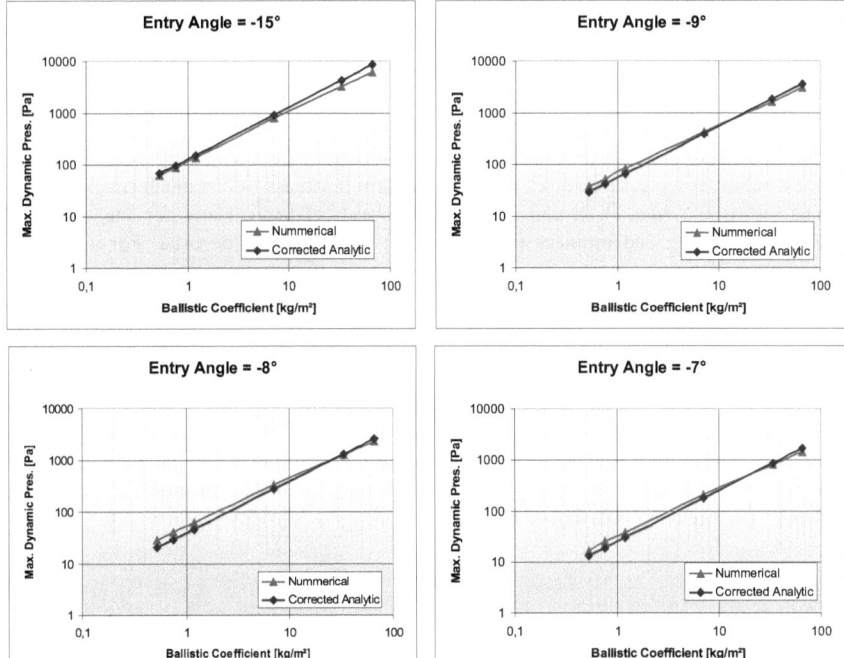

Fig. 7.14: Comparison between numerical and corrected analytical results for the maximum stagnation point pressure.

Naturally, maximum values can be replaced with values corresponding to a given design ballistic coefficient and from β_{design} then follows ϑ_{design} .

Note that in addition to this, the ballistic coefficient also has a significant impact on the descent speed, time and reachable maximum altitude. This circumstance is explored more thoroughly in a descend analysis given in chapter 7.2.3.

7.2.2   Determination of the Required Inflation Pressure Range

The upper limit of the ballute pressure is the skin stress which the envelope can comfortably handle with sufficient margin. The lower limit is determined by the ballute's desired deceleration characteristics.

Due to skin stress, interior overpressure will cause the ballute to expand slightly under the strain (see Fig. 7.15). The upper pressure limit is therefore given by the stress limit that the skin material and the seams can handle without failing or excessive plastic deformation. To find the upper pressure limit, we need to find a relation between the maximum pressure difference between interior and ambient pressure of the ballute and its skin and seam properties. The maximum pressure difference occurs most likely in space or around the point of maximum hypersonic heating where the gas temperature may be expected to be at its peak value for the mission. By definition of the task peak heating should occur at high altitudes, where the ambient pressure is so much smaller than the interior pressure, that it can safely be neglected for skin sizing:

$$\Delta p_{max} = p_{g,max} - p_{ambient,g,max}$$
$$p_{ambient,g,max} \ll p_{g,max} \qquad\qquad (89)$$
$$\Rightarrow \Delta p_{max} \sim p_{g,max}$$

Since a ballute is a pressure vessel, of which the skin thickness τ_s is much smaller than its radius r_b ($\tau_s \ll r_b$), it can be treated as such mathematically for the sake of a preliminary scaling and ordinary pressure vessel equations can be used. For a spherical pressure vessel, the skin stress σ_s is [54]:

$$\sigma_s = \frac{p_{g,max} \cdot r_b}{2\tau_s} \quad \text{in [MPa]} \qquad\qquad (90)$$

Note that the skin stress increases linearly with an increasing ballute radius, thus making a low ballistic coefficient (translates into a large but light structure) a quest for the strongest material available. In this equation, $p_{g,max}$ is the maximum expected ballute pressure and must not be confused with $p_{g,inf}$, the desired nominal ballute pressure at the time of inflation. Their relation is described by

$$p_{g,max} = p_{g,inf} \cdot \frac{T_{g,max}}{T_{g,inf}} \qquad\qquad (91)$$

$T_{g,max}$ is the maximum expected inflation gas temperature and must be obtained from a thermal analysis or a good estimate thereof. Alternatively, one might find it easier to define the absolute maximum pressure allowed in the ballute and use (91) to determine the required inflation pressure.

The maximum skin tension that the material can hold must always be greater than σ_s at $p_{g,max}$, so that we must demand

$$\sigma_{s,max} = \sigma_s \cdot \lambda_s \qquad (92)$$

with λ_s being a safety factor accounting for all uncertainties such as unequal stress distribution, vibration loads and others.

The definition of the lower pressure limit $p_{g,min}$ is somewhat more difficult to precisely define, since it needs to be high enough to prevent excessive deformation of the ballute. If the ballute pressure is too low, it will deform significantly during atmospheric entry, which is something we don't want. Deformation as such cannot be prevented of course, but it can well be maintained within acceptable limits. If the ballute changes its shape markedly, its aerodynamic properties change with it. Resulting changes in forces and moments will cause further deformation or, in a "wobbly" configuration, a shape change different to the one before, causing the ballute to oscillate in shape and orientation like a soap bubble until it disintegrates. Even if it stays intact, the trajectory of such an object is hardly desirable and definitely not predictable.

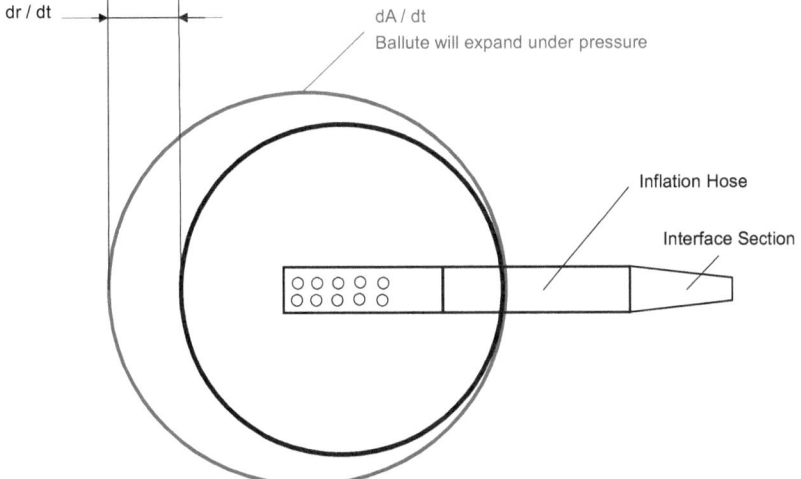

Fig. 7.15: Ballute expansion due to skin strain under pressure.

Equally important is the requirement that upper and lower pressure limits are far enough apart, so that the ballute can easily negotiate all external and internal forces that occur within its designated flight envelope. Since a sphere is the optimal solution to obtaining the maximum volume with the minimum surface area, any deformation of the ballute leads to a reduction in volume and consequently an increase in pressure. From equation (90) we see that the skin stress is directly proportional to the prevailing radius in a certain area. If the ballute departs from its spherical shape, some parts will decrease in radius and others will increase, further increasing local stress in that area. Consequently, if lower and upper pressure limits lie too close together, the balloon hull will fail, despite the fact that it would have held up out in space.

We must therefore see to it that the gas pressure of the ballute will maintain its shape within allowable limits, but stays below its allowable maximum with sufficient margin at all times. The precise shape determination of a flexible object subjected to the dynamic pressure of a flow field is subject to aeroelasticity studies and involves sophisticated CFD and FEM analyses. But even without these, we can attempt a reasonable estimate.

Two forces will cause the ballute to depart from its "unmolested" shape: one is the deceleration force acting on the instrument pod (during stable floatation also the pod's weight), the other is the pressure field of the flow field acting on the ballute's surface.

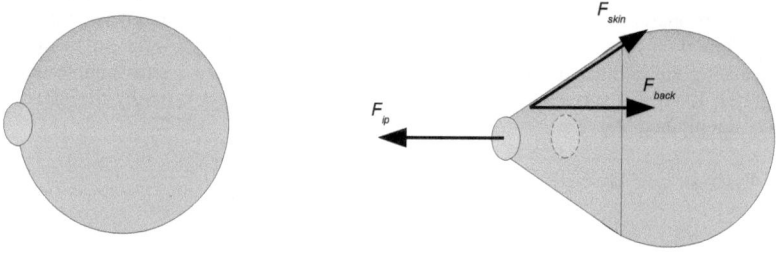

a) Spherical ballute in space   b) Spherical ballute with a forward load

Fig. 7.16: Shape change of a spherical ballute subjected to a forward load without additional load tendons or harnesses.

The pulling load is caused by the instrument pod pulling forward during deceleration in a stable attitude (centre of mass forward in flight direction) and the skin surrounding the instrument pod has to negotiate this load (details of sizing the fittings are given in chapter 7.2.5). Since film materials can only handle pulling loads (no compression loads, no out-of plane loads or moments) the ballute will deform until the skin surrounding the pod provides a force component pulling backwards which is large enough to counteract the pod's forward pull (see Fig. 7.16b). The shape change is therefore a function of the skin load, which is a function of the pressure difference between interior and exterior.

The pressure field close to the surface of the ballute immersed in a flow field (see Fig. 7.17) is caused by flow field particles impacting the skin surface (in case of a free particle stream) and the static pressure distribution resulting from the Bernoulli effect (in case of an inviscid continuum flow field). In the wake field, a low pressure zone will form that causes a local increase in the pressure difference acting on the skin, whereas around the stagnation point an overpressure zone will push the skin surface inward and therefore decrease the local pressure difference between inside and outside. The resulting shape change is again a function of the ballute's interior pressure.

Without much analysis, it becomes immediately clear that the higher the pressure difference between inside and outside the ballute is, the lesser pronounced its shape change will be. The possible skin stress, however, is limited by the skin material, so again the strongest material is the most desirable.

The simplest of approaches to an analytical analysis would regard the ballute as a fluid droplet, possessing a surface tension and an interior pressure, while immersed in a flow field exerting a shear stress. Such a situation is treated in reference [78]. A characteristic number (Weber number) is defined that sets into relation the surface tension and the flow field shear stress. If we now assume that the surface tension corresponds to the ballute's skin stress, the Weber number for the ballute can be written as:

$$We = \frac{\rho_{atm} v^2 d_{ballute}}{\sigma_s} \tag{93}$$

Since the skin stress of the ballute is a function of interior pressure and the shear stress is a function of the dynamic flow pressure, the Weber number for the ballute is also a relation between the dynamic pressure and the interior pressure.

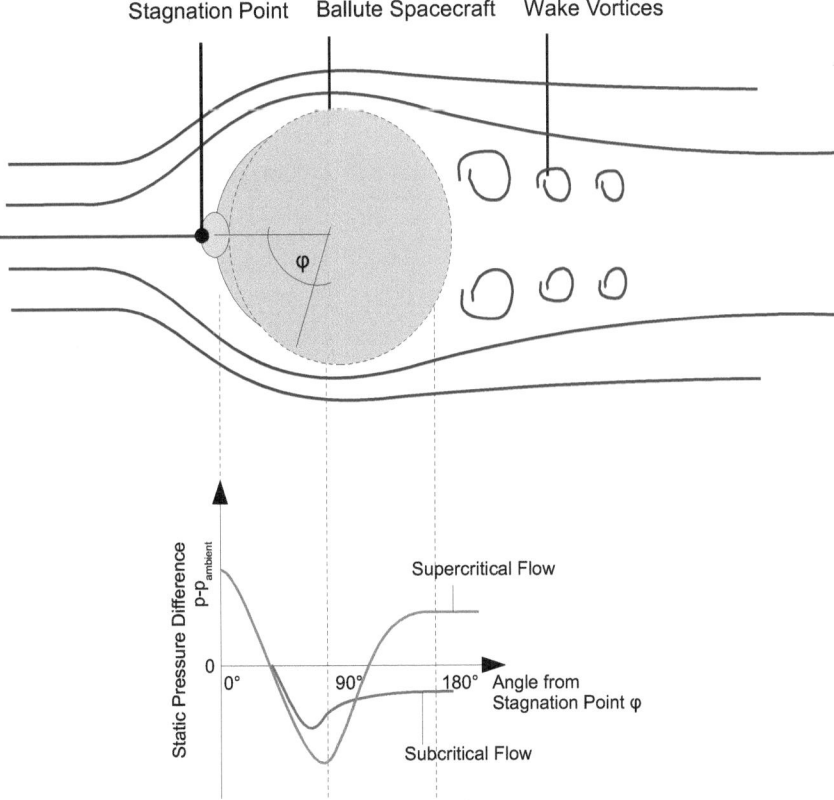

Fig. 7.17: Sketch of a typical continuum flow field around a spherical ballute with an Instrument-Pod-forward attitude. Shown in this sketch is a supercritical flow field with typical features and the static pressure distribution for subcritical and supercritical flows over the surface and the expected shape change resulting from it (based on reference [29]).

On any body immersed in a flow field, the stagnation point is defined as any point where the flow velocity is reduced to 0 and therefore the highest dynamic pressure occurs. For all studied ballute applications on Mars, maximum dynamic pressure stayed well below 50 Pa (in Earth entry cases below 100 Pa). On ARCHIMEDES this is well below 10% of the interior pressure, so the expected shape change will be minimal [78][79].

The highest stagnation point dynamic pressure occurs when the spacecraft reaches the trajectory point of maximum deceleration.

For the sake of envelope design we demand that

$$p_{g,min} > 10 \cdot q_{\infty,max} \tag{94}$$

for the ballute to maintain its shape within reasonable limits. For a preliminary study, the analytical method given in 7.2.1 may be used to determine the expected peak stagnation point pressure.

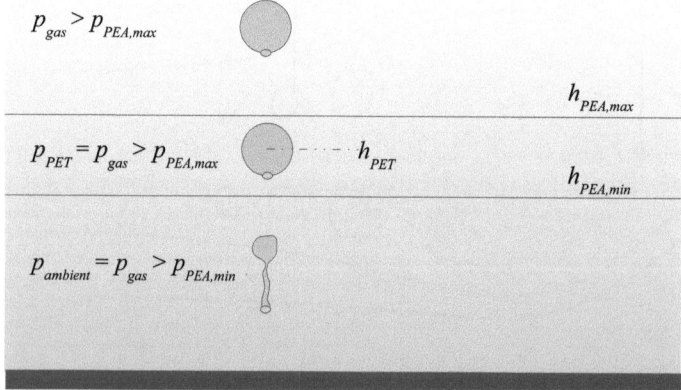

Fig. 7.18: Sketch of the pressure equilibrium transition event and altitude definition.

Last but not least, it must be decided at which altitude the ballute is allowed to collapse. In dense atmospheres, such as on Venus, full descent to the ground with a more or less constant velocity profile requires that the ballute collapses at a certain altitude to decrease its volume. On Mars, on the other hand, we might want to make sure that the ballute is still round and stiff when it touches the ground. But even on Mars, letting the ballute collapse at a predefined altitude might be desirable. Either way, we must make sure that the ballute's minimum interior pressure is greater than the ambient pressure at the lowest altitude we want the ballute to be fully rigid and round. We call this altitude "pressure equilibrium altitude" or PEA. The actual PEA depends on local weather conditions and the precise ballute pressure and temperature. The PEA band is therefore defined as the altitude range within which the pressure equilibrium transition event (PET event) occurs (see Fig. 7.18):

$$h_{PEA,min} < h_{PET} = h_{PEA} < h_{PEA,max} \tag{95}$$

The PEA range boundaries are a subject of mission design. The following conditions must be met by the inflation gas:

$$p_{g,min} > p_{\infty,PEA\,max} \quad \text{and} \quad p_{g,max} < p_{\infty,PEA\,min} \tag{96}$$

Care must be taken during mission design to make sure that the PEA range lies far enough below the sound barrier transition altitude. Because heavy vortex shedding and buffeting occurs at speeds between 2 Mach and 0.8 Mach, the ballute must be sufficiently pressurized in this regime to withstand the additional loads and vibrations.

7.2.3 Accessible Landing Terrain, Usable Altitude Range and Usable Descent Time

When designing a ballute mission, it may be of some importance to know the descent time of a given configuration to an altitude of interest, or even to tailor the spacecraft towards a desired descent time. Likewise, the accessible landing terrain might be of interest (the maximum allowable landing site elevation), or the maximum allowable ballistic coefficient if a soft touch down at a certain landing site with a given elevation is desired. In other words, we seek a method to design a spacecraft such that it has a predefined speed at a predefined altitude.

7.2.3.1 Transition Through Mach 1

Instruments exist that require the spacecraft to fly subsonic before they can start to operate. Additionally we may assume that landing should always occur at subsonic speeds. We therefore look at the flight phase from sound barrier transition onwards and start by finding an expression for the sound barrier transition altitude (SBTA).

The short interval between the altitude of maximum deceleration and the SBTA is the portion of the flight where the non-lifting spacecraft turns from its initial flight direction towards the ground (see Fig. 7.19).

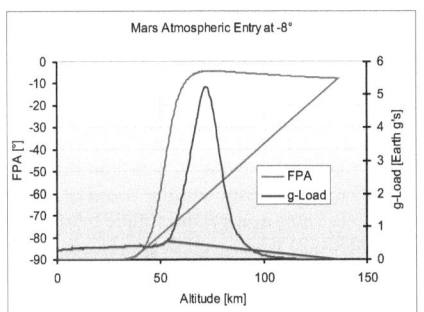
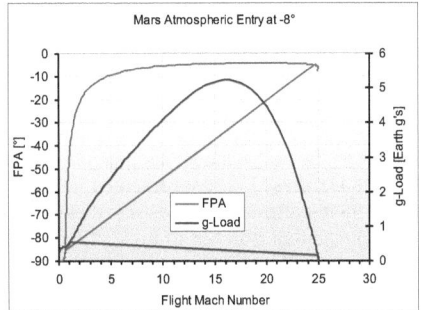

Fig. 7.19: Relation between sound barrier transition altitude and flight path angle for a shallow -8° entry into the Martian atmosphere of a vehicle with a ballistic coefficient of 0.5 kg/m².

To find the sound barrier transition altitude, we continue the trajectory from the point of maximum deceleration, modifying equation (68) such that it represents the spacecraft velocity, with respect to the point of maximum deceleration:

$$v_{sc} = v_{a,max} \exp\left[\vartheta \left(e^{\frac{-h_{a,max}}{H_0}} - e^{\frac{-h_{sc}}{H_0}}\right)\right] .$$ (97)

In (97), the values of the point of maximum deceleration have been used as new starting values. With the expressions for the maximum deceleration altitude and velocity (equations (71) and (86) respectively), (97) becomes:

$$v_{sc} = 0.6065 \, v_E \exp\left[\frac{1}{2} - \vartheta \, e^{\frac{-h_{sc}}{H_0}}\right]$$ (98)

We can solve (98) for the actual altitude and find the sound barrier transition altitude by substituting the speed of sound for the spacecraft velocity:

$$h_{Mach1} = H_0 \ln\left[-\frac{\vartheta}{\ln\left[\frac{v_{sound}}{0.6065 \, v_E}\right] - \frac{1}{2}}\right] \quad \text{with} \quad v_{sound} = \sqrt{\kappa_{atm} R_{atm} T_{atm}}$$ (99)

Temperature variations during a given time of day are small enough to yield acceptable results for a simple performance analysis. As discussed in chapter 7.2.1, this method is a good approximation only for comparatively steep entry angles, so again a correction factor for shallow angles has to be found. From chapter 3.4 and Fig. 3.17 we know that the impact of the flight path angle on the sound barrier transition altitude is minimal for entries more shallow than a certain limit angle and that in our case of a Mars entry from an elliptic orbit the limit is -10°. We therefore correct the entry angle accordingly:

$$\gamma = \begin{cases} \gamma_E & \text{for } -90° < \gamma_E < \gamma_{SBTA, limit} \\ \gamma_{SBTA, limit} & \text{for } \gamma_{SBTA, limit} \le \gamma_E \le \gamma_{shallow, limit} \end{cases}$$ (100)

for Mars entry at 4.5 km/s, $\gamma_{SBTA, limit} = -10°$ and $\gamma_{shallow, limit} = -7°$

Results of the above model compared to a numerical trajectory integration based on the Mars Climate Database [32] are presented in Fig. 7.20 and Fig. 7.21, assuming the speed of sound to be 190 m/s. Results demonstrate that deeper down in the atmosphere, the simple barometric altitude equation matched to high altitudes (from chapter 3.3) is not a good approximation of the actual profile any more, so using a different atmosphere layer with a different scale height is advised for heavier spacecraft with higher ballistic coefficients.

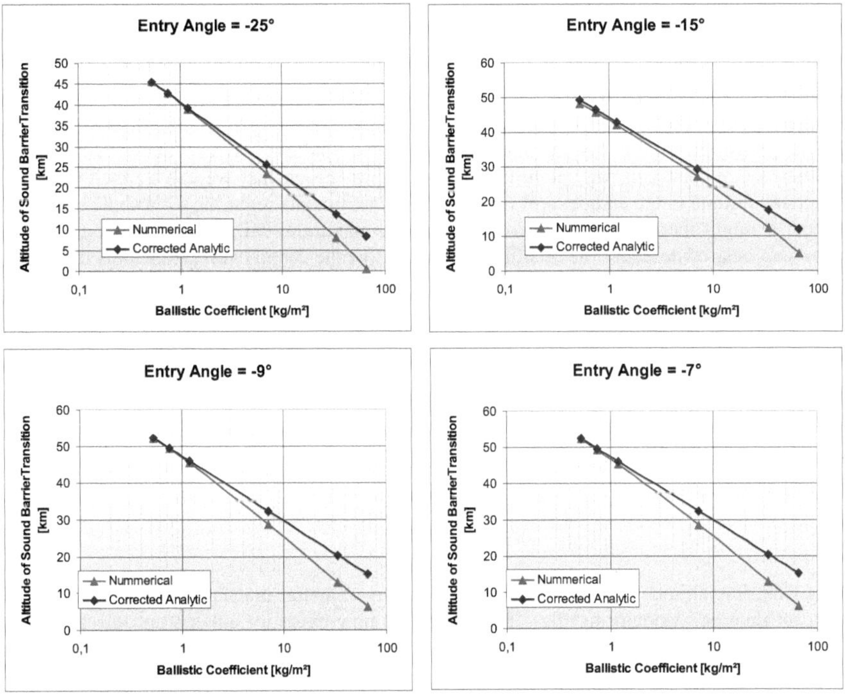

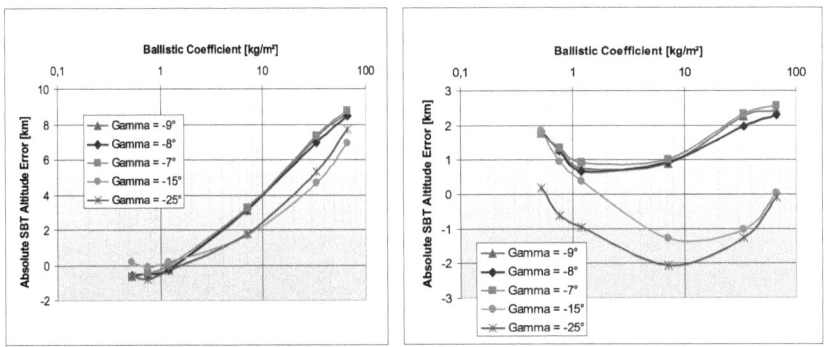

Fig. 7.20: Comparison of corrected analytic and numerical sound barrier transition altitude computations.

Fig. 7.21: Errors of the corrected analytic sound barrier transition altitude computation as compared to a numerical simulation based on an MCD profile [32]. Left with a barometric atmosphere model matched to high altitudes, right with one matched to lower altitudes.

However, since we may assume that the general altitude range of the intended mission profile is known or desired within a certain altitude band, the atmosphere model can be matched accordingly. Adapting the atmosphere to lower altitude profiles therefore improves results for higher ballistic coefficients with lower SBTAs (see Fig. 7.22).

Since we have already found out in chapter 3.4 that there is a limit for raising the sound barrier transition altitude by making the entry more shallow, we also know that only the ballistic coefficient can raise it further. This in turn means that we can herewith find the practical altitude range and descent time limit for instruments requiring a subsonic spacecraft speed. It might therefore be of interest to find the maximum ballistic coefficient allowed to achieve a desired descent time or altitude range bellow the sound barrier transition altitude.

We solve equation (98) for the ballistic coefficient, using equation (67) with $\gamma = \gamma_{SBTA,limit}$ and obtain an expression for the maximum allowable ballistic coefficient if the condition $h_{SBTA} \geq h_{SBTA,design}$ has to be met:

$$\beta_{max} = -\frac{H_0 \, \rho_0 \, e^{\frac{-h_{SBTA,design}}{H_0}}}{\left(2 \ln\left[\frac{v_{sound}}{0.6065 \, v_E}\right]\right) \sin\left(\gamma_{SBTA,limit}\right)} \tag{101}$$

It is from this altitude onwards that we wish to have a better understanding of the available descent time and velocity profile. Three scenarios now exist: we either want to reach stable floatation using the ballute as a super pressure balloon or we want to reach the ground of a certain elevation either with the ballute still super pressurized or collapsed.

Since the ballute must be comparatively rigid during sound barrier transition, we first look into a super pressurized descent.

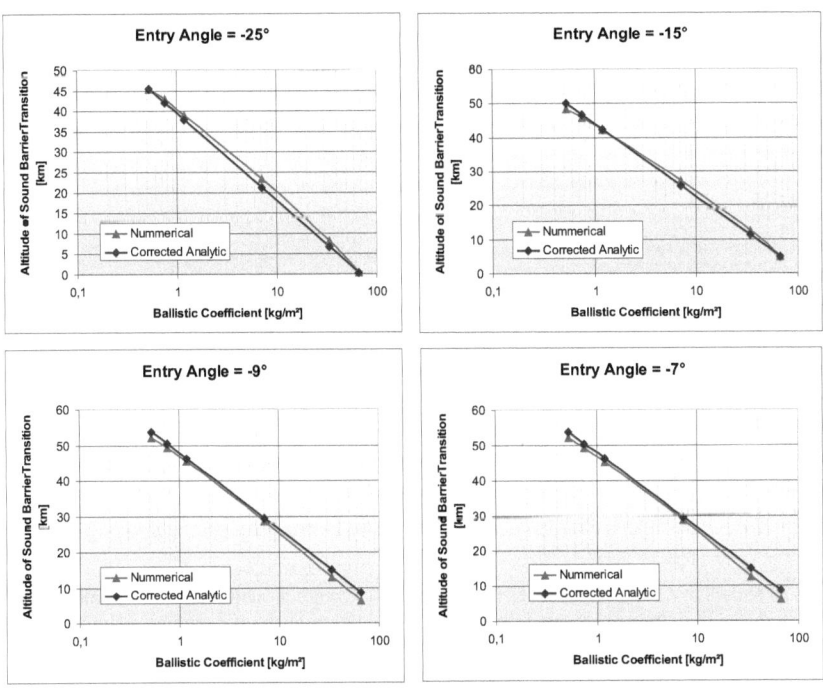

Fig. 7.22: Comparison of corrected analytic and numerical sound barrier transition altitude computations. Here a barometric altitude equation was used, which was matched to the lower Martian atmosphere.

7.2.3.2 Final Descent of a Super Pressurized Ballute

Taking the simple model discussed in chapter 7.2.1, we assume that after sound barrier transition the flight path angle is approximately 90° and the ballute descends straight down. The ballute is therefore in a stable equilibrium of all forces, its weight being compensated by aerodynamic drag and static lift. We rewrite the basic equations of motion (equations (4) to (9)) accordingly and obtain:

$$m_{sc} g = \frac{1}{2} \rho_\infty v_{sc}^2 A_{sc} c_d + \rho_\infty V_{BL} g \qquad (102)$$

If the ballute is purely spherical, we can find the following expression for the ballute volume as a function of the ballistic coefficient:

$$V_{BL} = \frac{4\, m_{sc}\, r_b}{3\, \beta\, c_d} \qquad (103)$$

With this equation and our simple barometric atmosphere model, (102) can be rewritten to yield the spacecraft velocity in terms of the ballistic coefficient and altitude:

$$v_{sc}^2 = \frac{2\,g\left(\beta\,e^{\frac{h_{sc}}{H_0}} - \frac{2}{3}\frac{r_b}{c_d}\rho_0\right)}{\rho_0} \tag{104}$$

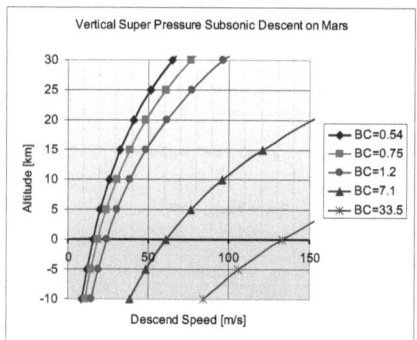 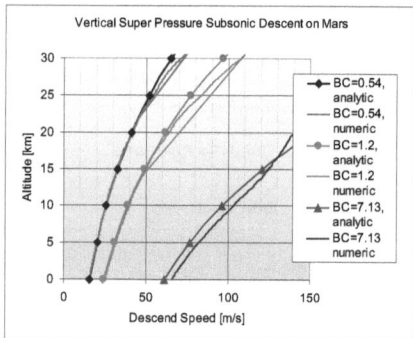

Fig. 7.23: Descent velocity for a vertical super pressurized descent on Mars for various ballistic coefficients (left) and compared to a numerical simulation based on the MCD.

Fig. 7.23 shows that the analytical results are in good agreement with numerical computations for as long as the atmosphere models are a good match and the descending vehicle is well below the sound barrier.

To further simplify matters, we introduce the following parameter:

$$\xi = \frac{2}{3}\frac{r_b}{c_d}\rho_0 \tag{105}$$

This parameter represents the "lifting" force and may be regarded as the "equivalent ballistic coefficient" of the displaced atmosphere gas. We can see that the term becomes relevant only if the ballute radius is significantly large or else the term will be very small.

Let the landing terrain have 0m elevation and the touch-down speed can be obtained by

$$v_{TD} = \sqrt{\frac{2\,g\,(\beta-\xi)}{\rho_0}} \tag{106}$$

If now a ballistic coefficient is sought to provide a certain speed at a certain altitude, for example a a maximum landing speed on an interesting plateau on Mars, we may solve equation (104) for a design ballistic coefficient:

$$\beta_{design} = \left(v_{design}^2\frac{\rho_0}{2\,g} + \xi\right)\exp\left[\frac{-h_{design}}{H_0}\right] \tag{107}$$

To find the descent time from a certain altitude A below the SBTA to another altitude B, equation (104) can be rewritten with $v_{sc} = -\dfrac{dh}{dt}$ to yield the descent time:

$$dt = \sqrt{\frac{\rho_0}{2g}} \cdot \frac{1}{\sqrt{\beta \, e^{\frac{h_{sc}}{H_0}} - \xi}} \, dh \qquad (108)$$

Integrating the above equation from point A to point B yields

$$\Delta t_{A-B} = 2 H_0 \sqrt{\frac{\rho_0}{2g\xi}} \cdot \left[\arctan\left(\sqrt{\frac{\beta \, e^{\frac{h_A}{H_0}} - \xi}{\xi}} \right) - \arctan\left(\sqrt{\frac{\beta \, e^{\frac{h_B}{H_0}} - \xi}{\xi}} \right) \right] \qquad (109)$$

The evaluation of equation (109) for the descent of a ballute to the surface of Mars is shown and compared to numerical analysis in Fig. 7.24. The atmosphere model used was the standard model adequate for the lower 30km of the Martian atmosphere, therefore numerical results differ slightly for altitudes above that.

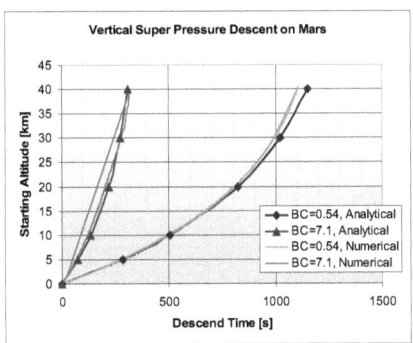
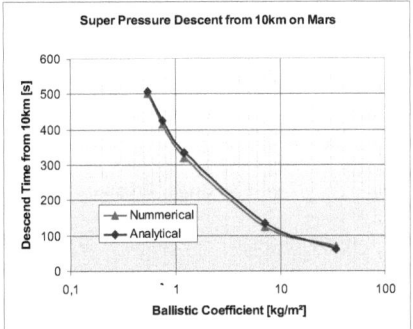
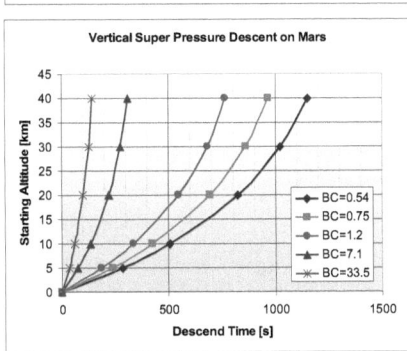
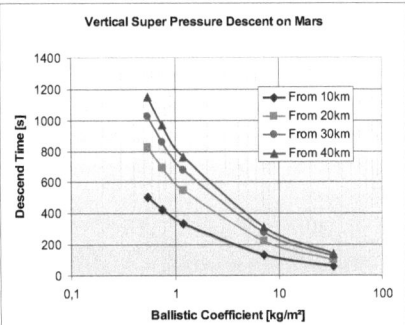

Fig. 7.24: Descent time of a spherical super pressure ballute on Mars.

Equation (109) can also be derived in terms of the spacecraft mass and volume, for an arbitrary vehicle:

$$\Delta t_{A-B} = 2 H_0 \sqrt{\frac{A_{SC} c_D}{2 g V_{BL}}} \cdot \left[\arctan\left(\sqrt{\frac{m_{SC} e^{\frac{h_A}{H_0}} - V_{BL} \rho_0}{V_{BL} \rho_0}} \right) - \arctan\left(\sqrt{\frac{m_{SC} e^{\frac{h_B}{H_0}} - V_{BL} \rho_0}{V_{BL} \rho_0}} \right) \right] \quad (110)$$

Depending on the desired analysis, A can now be the sound barrier transition altitude from equation (99) and B a target altitude, for instance the ground or the pressure equilibrium altitude. If B is the ground at altitude 0, equation (109) becomes:

$$\Delta t_{Mal-Gnd} =$$

$$2 H_0 \sqrt{\frac{\rho_0}{2 g \xi}} \cdot \left[atan\left(\sqrt{\frac{-\dfrac{H_0 \rho_0}{2 \sin(\gamma)\left(\ln\left[\dfrac{v_{sound}}{0.6065\, v_E}\right] - \dfrac{1}{2} \right)} - \xi}{\xi}} \right) - atan\left(\sqrt{\frac{\beta - \xi}{\xi}} \right) \right] \quad (111)$$

The descent time from SBTA to ground with respect to the entry angle is depicted in Fig. 7.25.

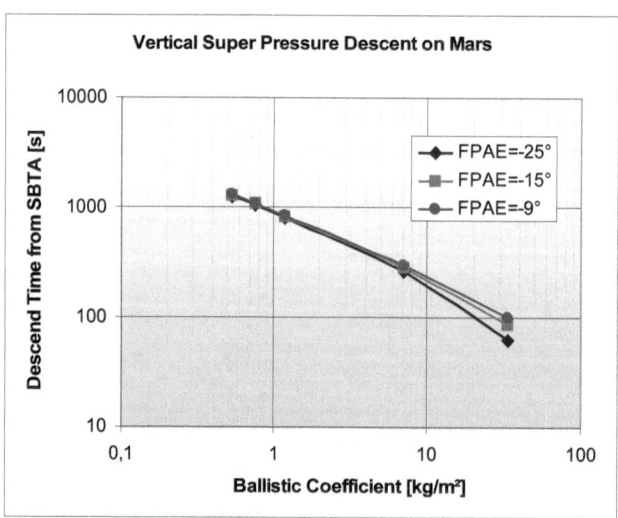

Fig. 7.25: Descent time from sound barrier transition to the ground (elevation = 0m) with respect to the ballistic coefficient for various entry angles on Mars.

7.2.3.3   Final Descent of a Collapsing Ballute

This last mission phase will not be studied in much detail for two reasons: for one, the descent of a collapsing ballute cannot easily be modelled, since the large and flexible skin provides no reliable prediction of the drag and momentum coefficients, as it heavily interacts with the airflow and for another, it is not within the prime scope of this study. For good measure, however, it should be included, as it delivers acceptable results for principle parameter studies and a collapsing ballute is a viable mission design method. The Mars mission concept ARCHIMEDES (see chapter 10.3) does not rely on a collapsing ballute phase, but the flight test MIRIAM did (see chapter 10.2), as well as a mission study for Venus (ARCHIMEDES-V, see chapter 10.4.1).

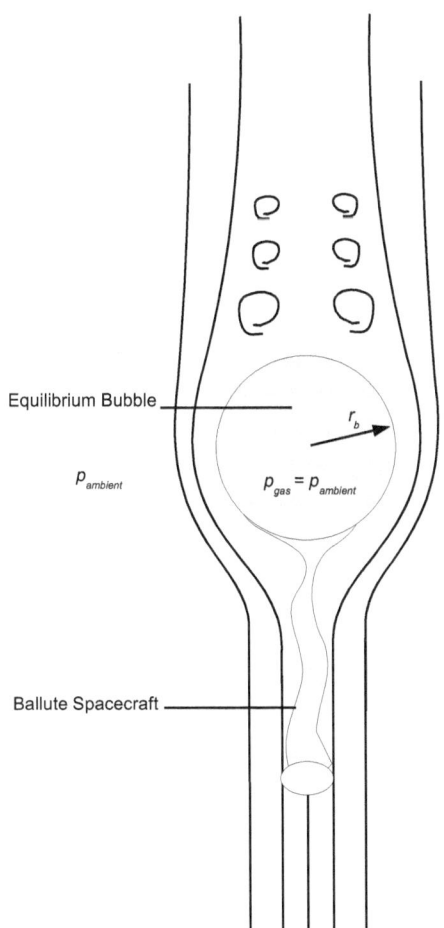

Fig. 7.26: Sketch of the collapsing ballute during descent.

A mission phase with a collapsing ballute becomes relevant when continued descent into thicker atmospheres is required. A low ballistic coefficient might prevent such a descent altogether, yet might still be desirable for a high altitude deceleration profile. In other words, for hypersonic flight a ballistic coefficient so low might be desired that stable floatation would lie above the target altitude. In such a case we may choose a variable ballistic coefficient, which can either be achieved by jettisoning equipment (i.e. jettisoning the ballute and replacing it with a smaller one or a parachute or something completely different) or by using a ballute of variable size (i.e. by allowing its collapse under external pressure during descend).

This latter scenario has the advantage of working without the requirement for additional moving parts or actuators, but at the expense of a less predictable trajectory (as stated above, the flexible skin has aerodynamic properties that are hard to model or predict). But for an atmospheric sounding probe, that doesn't really matter much. Such a scenario was actually studied for the flight test MIRIAM (see Chapter 10.2) and a possible atmospheric sounding ballute for Venus (see Chapter 10.4.1), where continuous measurements from the outer atmosphere layers down to the ground was required.

Further to the assumptions we have already made, we treat the ballute as if it shrank in size, such that its characteristic parameters match those of a spherical gas bubble in pressure equilibrium. In other words, we assume the ballute to be fashioned from an elastic skin, like the one of a toy balloon, assuming no contribution of the skin tension to the interior pressure. This assumption is justifiable for ballutes filled with a buoyant gas, which collects in a bubble on top of the ballute and leaves the rest of the skin hanging loose.

We define the term "equilibrium bubble" (see Fig. 8.5) as the spherical volume of the inflation gas under pressure equilibrium. Any part of the ballute skin that does not contribute to the shape of the equilibrium bubble is neither considered to contribute much to the drag nor the volume of the ballute any more. If the descend velocity is low enough, this assumption is crude, yet acceptable.

We start again by assuming an equilibrium of all forces during final descend as given in equation (102). This leads to the following expression for the descend velocity:

$$v_{sc}^2 = \frac{m_{sc}g - \rho_\infty V_{BL}g}{\frac{1}{2}\rho_\infty A_{sc}c_D}$$

(112)

Note that due to the collapsing ballute, the volume and face area are now functions of the instantaneous altitude. From the assumption of the equilibrium bubble follows

$$p_g \approx p_\infty$$

(113)

and with it the instantaneous ballute volume

$$V_{BLi} = \frac{m_g R_g T_g}{p_0 e^{\frac{-h_{sc}}{H_0}}}$$

(114)

and radius

$$r_b^3 = \frac{3}{4} \frac{V_{BLi}}{\pi} \ . \tag{115}$$

Because of equation (113), the density of the ballute inflation gas at the "bottom" of the atmosphere (where $p_{infinite} = p_0$ and h = 0m) is

$$\rho_{g,0} = \frac{p_0}{R_g T_g} \tag{116}$$

If a collapsing ballute is required, the mass of the inflation gas has to be determined through the desired pressure equilibrium transition altitude h_{PET}. Also because of equation (113), it can be determined by the following expression:

$$m_g = \frac{V_{BL} \, p_0}{R_g T_g} e^{\frac{-h_{PET}}{H_0}} = V_{BL} \, \rho_{g,0} \, e^{\frac{-h_{PET}}{H_0}} \tag{117}$$

Note that in the above two equations, a good estimate of the inflation gas temperature at the point of pressure equilibrium is required!

Equation (112) can now be rewritten as a function of the instantaneous altitude by replacing ambient density, ballute volume and spacecraft face area with the appropriate terms of the equilibrium bubble:

$$v_{sc}^2 = \frac{2g\left(m_{sc} - \rho_0 V_{BL} e^{-\frac{h_{PET}}{H_0}}\right)}{\rho_0 c_D \pi e^{-\frac{h_{sc}}{H_0}} \left(\frac{3}{4\pi} V_{BL} e^{\frac{h_{sc} - h_{PET}}{H_0}}\right)^{\frac{2}{3}}} \tag{118}$$

For the mass of the ballute spacecraft, we can use the general mass budget equation (1) (chapter 3.1.1 on page 16) and replace the gas mass by equation (117) to once again rewrite equation (112), this time even taking the gas mass of the equilibrium bubble into account:

$$v_{sc}^2 = \frac{2g\left(m_{ip} + m_H + V_{BL} e^{-\frac{h_{PET}}{H_0}}\left(\rho_{g,0} - \rho_0\right)\right)}{\rho_0 c_D \pi e^{-\frac{h_{sc}}{H_0}} \left(\frac{3}{4\pi} V_{BL} e^{\frac{h_{sc} - h_{PET}}{H_0}}\right)^{\frac{2}{3}}} \tag{119}$$

To simplify further treatment of the matter, we define a "weight parameter" that includes all parameters that are constant (or assumed to be constant):

$$\zeta = \frac{2g}{\pi \rho_0 c_D \left(\frac{3}{4\pi}\right)^{\frac{2}{3}}} \left[m_{ip} + m_H + V_{BL} e^{-\frac{h_{PET}}{H_0}}\left(\rho_{g,0} - \rho_0\right)\right] \quad \text{in} \quad \left[\frac{kg \, m}{s^2}\right] = [N] \tag{120}$$

This parameter describes the force with which the collapsing ballute spacecraft is pushing down on the surrounding atmosphere. It is referenced to a datum altitude of h=0 (through the term $\left(\rho_{g,0} - \rho_0\right)$). The weight parameter can be reduced by using an inflation gas of lesser density than the atmosphere gas.

As stated before, the Mars ballute ARCHIMEDES is intended to reach the surface of the planet fully inflated, but the equations herein were used in the evaluation and design of the

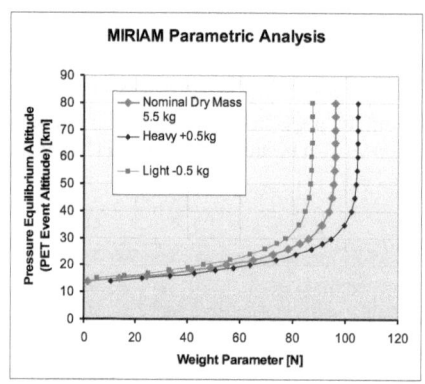
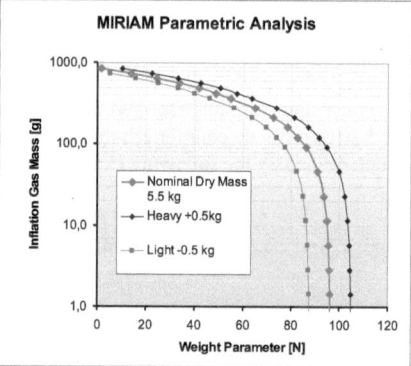

Fig. 7.27: Weight parameter analysis from the MIRIAM system study. Left the weight parameter as a function of the PEA and right the required inflation gas mass as a function of required weight parameter.

ballute spacecraft of flight test MIRIAM, which was supposed to collapse at a certain altitude (see Chapter 10.2).

Fig. 7.27 shows the weight parameter for nominal, heavy and light configurations, as a function of the PET event altitude PEA and the required inflation gas mass of the MIRIAM ballute spacecraft. MIRIAM had a 4-m UPILEX-25RN ballute with a production factor of 1.4 and was filled with Helium (see Chapter 10.2).

For the case of a vehicle coming to rest exactly the moment it touches the ground at h=0, we can find the corresponding minimum PEA:

$$h_{PET,min} = H_0 \ln\left(\frac{-V_{BL}\left(\rho_{g,0} - \rho_0\right)}{m_{ip} + m_H}\right) \tag{121}$$

Note that in (116) the assumed ballute gas temperature at the surface plays a role, but that the expression for the weight parameter has (116) as part of a density difference on the surface. The expected error from a bad temperature estimate is therefore minimal.

Fig. 7.28 shows the influence of the inflation gas temperature at zero altitude on the pressure equilibrium altitude, where PET occurs and the weight parameter. While this is minimal, it should be kept in mind that the entire model is based on the assumption of an isothermal model atmosphere. Therefore, the real instantaneous gas temperature does play a role in real life and can vary with daytime, season and weather conditions at the time of descend.

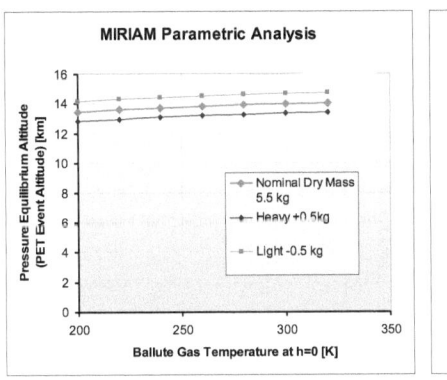 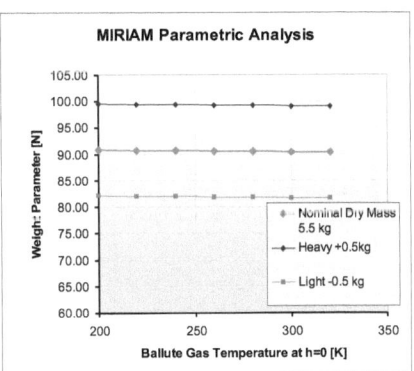

Fig. 7.28: Dependence of the PEA (left) and the weight parameter (right) on the temperature of the ballute gas.

This brings us to the final version of equation (112), which now becomes

$$v_{sc}^2 = \cfrac{\zeta}{\rho_0 \exp\left[-\cfrac{h_{sc}}{H_0}\right]\left[V_{BL} \exp\left[\cfrac{h_{sc}-h_{PET}}{H_0}\right]\right]^{\frac{2}{3}}} \tag{122}$$

The touch-down velocity of a collapsing ballute can be found by setting $h_{sc}=h_{TD}$. If the terrain elevation is 0 (touch down at mean sea level or in more general terms, the atmospheric reference datum), the speed is

$$v_{TD} = \sqrt{\cfrac{\zeta}{\rho_0 \left(V_{BL} e^{\frac{-h_{PET}}{H_0}}\right)^{\frac{2}{3}}}} \tag{123}$$

The above equation shows nicely that the touch down velocity increases with the weight parameter (a heavy spacecraft or a heavy planet with a high surface acceleration). It is low for dense atmospheres and high for thinner atmospheres. The touch down velocity also increases when the pressure equilibrium altitude h_{PET} is raised (see Fig. 7.29). This is obvious as the ballute collapses earlier and therefore displaces less atmosphere gas and has less dynamic resistance in flow field.

The touch down speed is zero if the weight parameter is. Equation (120) shows that the weight parameter becomes zero if the term in brackets gets zero, in other words, if the atmosphere gas displacement just so negates the vehicle's dry weight and the vehicle starts to float the moment it reaches the ground. If the weight parameter gets negative, the vehicle will not descend to the ground, but remain floating at an altitude for which a weight parameter referenced there would be zero.

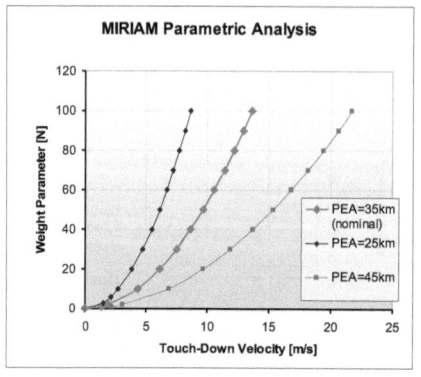 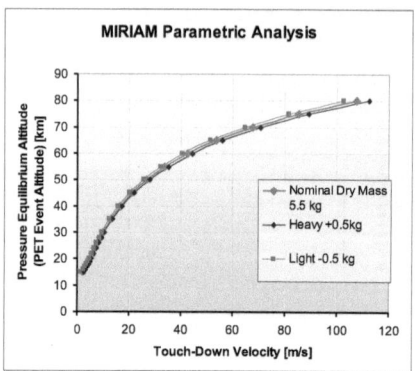

Fig. 7.29: From the MIRIAM parametric study: Left and right the touch down velocity as a function of the Weight Parameter and the PEA. Shown are nominal, heavy and light configurations for a 4-m UPILEX-25RN ballute with a production factor of 1.4.

To find the descent time from the PET event (in other words, the pressure equilibrium altitude PEA) to a certain target altitude B, we rearrange the above with

$$v_{sc} = -\frac{dh_{sc}}{dt} \tag{124}$$

to obtain

$$dt = \sqrt{\frac{\rho_0}{\zeta}} \sqrt[3]{V_{BL}} \, e^{-\frac{h_{sc}+2h_{PET}}{6H_0}} dh_{sc} \tag{125}$$

and carry out the required integration from PET to B to arrive at an expression for the descent time:

$$\Delta t_{PET-B} = \sqrt{\frac{\rho_0}{\zeta}} \sqrt[3]{V_{BL}} \, 6H_0 \, e^{-\frac{h_{PET}}{2H_0}} \left(1 - e^{\frac{h_B - h_{PET}}{6H_0}}\right) \tag{126}$$

For the sake of a quick discussion, we observe that if the target altitude lies close to the PET altitude, the descent time is short. In case they are identical, the descent time is 0. The descent time is shortened by a large weight parameter, increases with ballute volume and a thicker and heavier atmosphere. Note that the descent time increases with the altitude difference to the power of e. Results for a parametric analysis of MIRIAM are given in Fig. 7.30.

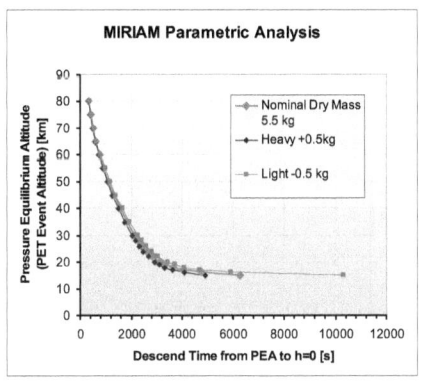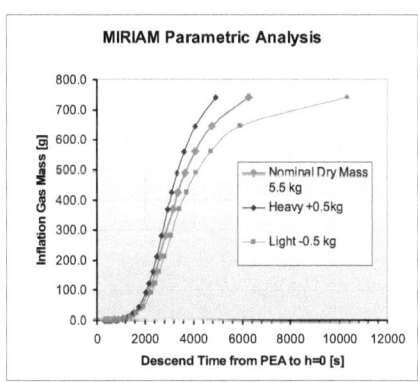

Fig. 7.30: Study of the descent time from the PEA as a function of the PEA and the required inflation gas mass for MIRIAM.

The required gas mass for a desired PEA can be found quite simply through the ideal gas law, because the PEA is the point where the ballute still has its full volume, but the inflation gas pressure is already the same as the ambient pressure:

$$m_{g,design} = \frac{p_0 V_{BL}}{R_g T_g} e^{-\frac{h_{PET}}{H_0}} \qquad (127)$$

With (121) we obtain the inflation gas mass required to achieve a touch-down velocity of exactly zero:

$$m_{g,design,v_{TD}=0} = \frac{-p_0 V_{BL}^2 (\rho_{g,0} - \rho_0)}{R_g T_g (m_{ip} + m_H)} \qquad (128)$$

Note, however, that this relation is quite sensitive to unknowns, such as gas temperature and real world weather conditions and therefore should be understood as a limit value, from which a sufficient clearance must be maintained, if reaching the ground is a mission objective.

In Fig. 7.31, results from the above derived velocity computations are compared against a numerical trajectory simulation for MIRIAM's collapsing ballute. PET for MIRIAM in the studied scenario occurs at 34.4km. The analytical results are in good agreement with the numerical results, although the latter also take the filling gas and atmosphere temperatures into account. The simple barometric atmosphere model used above is explained in chapter 3.3 and depicted in Fig. 3.6.

Note again that the above theory is based on the simplification that the ballute assumes a shape which makes those parts of the skin with no skin tension contribute negligibly little to the overall drag.

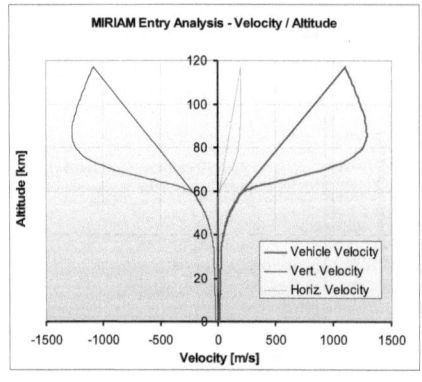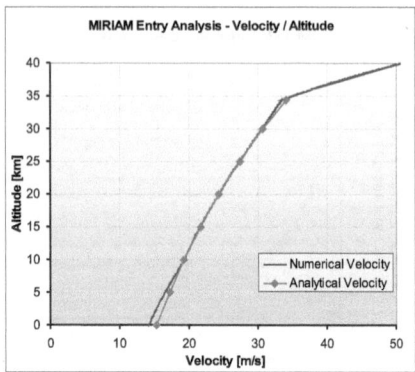

Fig. 7.31: Numerical descent velocity computation for the MIRIAM ballute spacecraft, coming in on a ballistic trajectory with a 180km peak altitude (left) and final descent of the collapsing MIRIAM ballute spacecraft, compared to analytical results for the same mission (right).

7.2.4   Sizing the Ballute

In equation (1) (chapter 3.1.1 on page 16), the ballute spacecraft mass is defined as the sum of the instrument pod and ballute masses:

$$m_{SC} = m_{BL} + m_{ip} \tag{129}$$

where the ballute mass is the sum of the hull mass and the inflation gas mass:

$$m_{BL} = m_H + m_{gas} \tag{130}$$

The hull mass in turn is the sum of the masses of a perfect thin film sphere and the real world production mass fraction. The latter may also be expressed as the mass difference between the real world ballute after its completion and a perfectly seamless sphere of equal dimensions (skin thickness and diameter). The hull may therefore be expressed as

$$m_H = m_s + m_{pf} \tag{131}$$

There is no practical analytical way of determining m_{pf} , since it accounts for everything not belonging to a perfect sphere. The production mass fraction therefore comprises all seams, reinforcements and attachment points, the inflation hose and windsock and instrumentation possibly incorporated in the ballute skin. These masses are all highly specific to the ballute's production process and intended use, so for a parametric analysis, it is appropriate to introduce an empirically determined production factor, that relates the production mass fraction to the skin mass of the dimensionally equivalent sphere:

$$m_{pf} = \lambda_{pf} m_s \tag{132}$$

Typical values of λ_{pf} achieved during the ARCHIMEDES development programme range between 0.1 and 0.4 for a simple, welded Polyethylene balloon and a fully instrumented 4 metre Polyimide ballute, respectively. With (132), equation (131) becomes

$$m_H = m_s(1+\lambda_{pf}) \tag{133}$$

The mass of the dimensionally equivalent sphere may be obtained by

$$m_s = \varrho_A{}' \cdot A_H \tag{134}$$

where $\varrho_A{}'$ is the skin material's areal density in $\left[\dfrac{kg}{m^2}\right]$, a property value of fabric and film

materials. Typical values for everyday thin films range between $1\dfrac{g}{m^2}$ for very light and thin

materials such as household cling film (or plastic wrap) and $100\dfrac{g}{m^2}$ for a sturdy pond liner.

Typical values for fabrics range from $10\dfrac{g}{m^2}$ for very thin silk to as much as several

$1000\dfrac{g}{m^2}$ for multi-layer protective clothing, such as the ones used by fire fighters. Typical

high altitude balloons have an areal skin density range between $6\dfrac{g}{m^2}$ and $30\dfrac{g}{m^2}$. Now

these values are anything but exact, but should give the reader a ball-park figure of typical materials.

With (134) and (57), the hull mass can be obtained from all characteristic parameters, by rewriting equation (131) to

$$m_H = \varrho_A{}' 4\pi r_b^2 (1+\lambda_{pf}) \tag{135}$$

For all relevant ballute pressures, we may safely assume ideal gas behaviour. The gas mass is therefore given by

$$m_{gas} = \frac{P_g V_{BL}}{R_g T_g} = \frac{P_{g.inf}}{R_g T_{g.inf}} \cdot \frac{4}{3}\pi r_b^3 \tag{136}$$

Note that in (136), the gas temperature $T_{g.inf}$ and pressure $P_{g.inf}$ are the values at the time when the ballute was pressurized, as they determine the amount of gas actually flowing into the ballute.

Assuming that we know the available materials, inflation gasses and the desired payload (and therefore Instrument Pod mass), the ballistic coefficient of the configuration can be found by combining equations (129), (135) and (136) with equation (16):

$$\beta = \frac{m_{ip} + \varrho_A{}' 4\pi r_b^2 (1+\lambda_{pf}) + \dfrac{P_{g.inf}}{R_g T_{g.inf}} \cdot \dfrac{4}{3}\pi r_b^3}{c_d \pi r_b^2} \tag{137}$$

It is more likely, however, that the ballistic coefficient is fixed, because it is so desired (from a scientifically required mission profile, see chapter 7.2.1). The available areal density of the ballute skin is also likely to be known down to a certain number of options, as available films only come in certain gauges, especially high performance thin films. A trade-off must therefore be made between Instrument Pod mass and a convenient ballute radius. The latter is usually limited by available packing volume and the former driven by the payload wish-list. The relation between the two is given by rewriting (137) as

$$m_{ip} = \left[c_d \beta - 4 \varrho_A{}' (1+\lambda_{pf}) \right] \pi r_b^2 - \frac{p_{g,inf}}{R_g T_{g,inf}} \cdot \frac{4}{3} \pi r_b^3 \qquad (138)$$

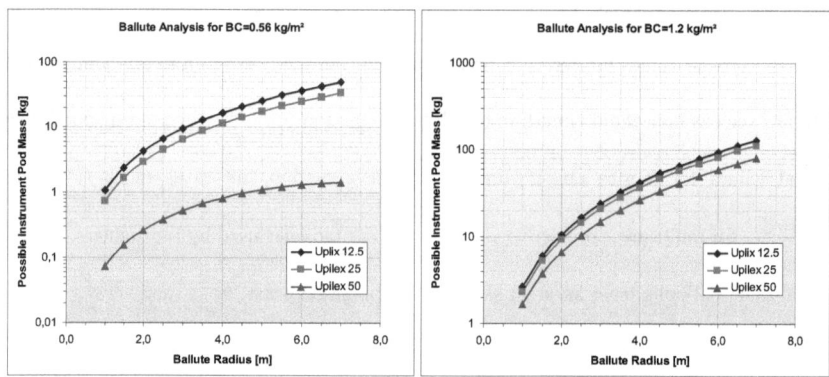

Fig. 7.32: Possible Instrument Pod mass as a function of ballute radius for various UPILEX films and desired ballistic coefficients (without considering material strength).

The possible Instrument Pod mass as a function of the ballute radius is given in Fig. 7.32. A production factor 0.4 (as it was achieved for the fully instrumented MIRIAM ballute) was assumed for this study. Note that in equation (138), only the areal density is used and nothing is said here about the material strength and load. The relation between the areal density and the skin stress is discussed further on in this chapter. We can already see, however, that there is an achievable limit ballistic coefficient for a ballute without an Instrument Pod, which only depends on the available material gauge and possible production factor. This can be easily found, by setting the Instrument Pod mass to zero:

$$\beta_{limit} = \frac{\varrho_A{}' 4 \pi r_b^2 (1+\lambda_{pf}) + \frac{p_{g,inf}}{R_g T_{g,inf}} \cdot \frac{4}{3} \pi r_b^3}{c_d \pi r_b^2} \qquad (139)$$

The difference between the desired ballistic coefficient and the achievable limit ballistic coefficient is the margin that can be used for a payload carrying Instrument Pod. The limit ballistic coefficient is shown in Fig. 7.33, for selected UPILEX thin film variants and production factors. It has only a comparably weak dependence on the ballute radius, but we can already see that improving the production factor through better techniques and lowering the skin density, by using a higher strength thin film can greatly improve the ballute performance.

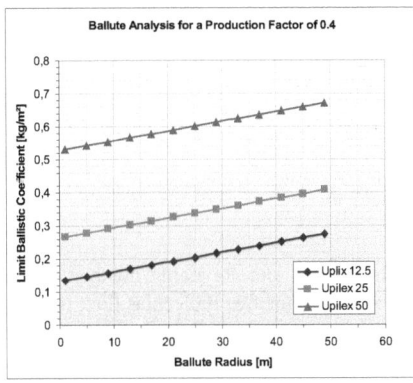 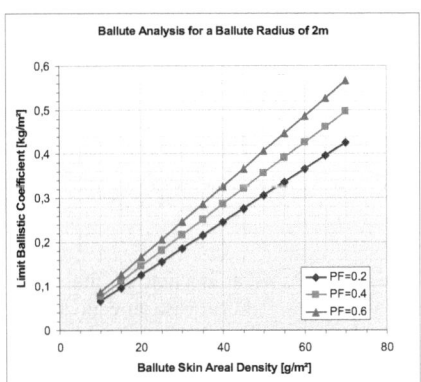

Fig. 7.33: Analysis of the limit ballistic coefficient for selected skin materials and production factors.

If the ballute radius needs to be obtained from (138) as a function of the Instrument Pod mass, this is best done by Newton iteration. Although an exact analytical solution exists, it is a rather large and as such not a very practical expression.

In some cases very light gasses such as Helium or Hydrogen may be considered, so that $m_{ip} + m_H \gg m_{gas}$ is true and

$$\beta \approx \frac{m_{ip} + m_H}{c_d A_{SC}} \tag{140}$$

In this case, the resulting ballute radius may be obtained from

$$r_b = \sqrt{\frac{m_{ip}}{\left[c_d \beta - 4 \varrho_A{}' (1 + \lambda_{pf}) \right] \pi}} \quad \text{in [m]} \tag{141}$$

It is interesting to further discuss the term in the denominator of (141), because it contains the desired ballistic coefficient (read "mission aspirations") and the actual mean areal skin density over the entire ballute (read "technical capabilities"). Because $c_d \beta = \dfrac{m_{SC}}{A_{SC}}$, this is the difference between the aerodynamically effective areal density of the spacecraft and the actual areal density of the ballute envelope. The areal density of the spacecraft is thereby reduced to the desired value by an increase in radius, which of course only works, if the mean areal density of the ballute is "lighter" than that because the Instrument Pod is so much denser (this leads us back to the achievable limit ballistic coefficient, already discussed above). In other words: the smaller this difference, the larger the inner volume has to be to "dilute" the Instrument Pod mass. Therefore the performance of the ballute is directly related to the performance of the hull (read: skin material properties and production techniques). The theoretical limit is obviously

$$c_d \beta - 4 \varrho_A{}' (1 + \lambda_{pf}) = 0 \tag{142}$$

where the average skin density itself approaches the desired areal density of the spacecraft and, therefore, the resulting ballute radius has to increase towards infinity.

Because (142) contains no expression for the inflation gas, it is equivalent to a body containing vacuum. So for a parametric analysis, we may define an independent figure of Merit for spherical ballutes:

$$B_{M,Sphere} = c_d \beta - 4\varrho_A{}' (1+\lambda_{pf}) \quad \text{in} \quad \left[\frac{kg}{m^2}\right] \tag{143}$$

Note that c_d is a function of the flight Mach number and Reynolds number of the prevailing flow field (in case an equivalent drag coefficient for free molecular flow is used) and is therefore not easily obtained. If we do not need to express the ballute figure of merit as a function of the ballistic coefficient, we might therefore prefer the following expression:

$$B_{M,Sphere} = \frac{m_{sc}}{A_{sc}} - 4\varrho_A{}' (1+\lambda_{pf}) \tag{144}$$

Either way, expression (143) for the ballute figure of merit contains all the information which characterises the mission and available technical capabilities pertaining to the ballute body.

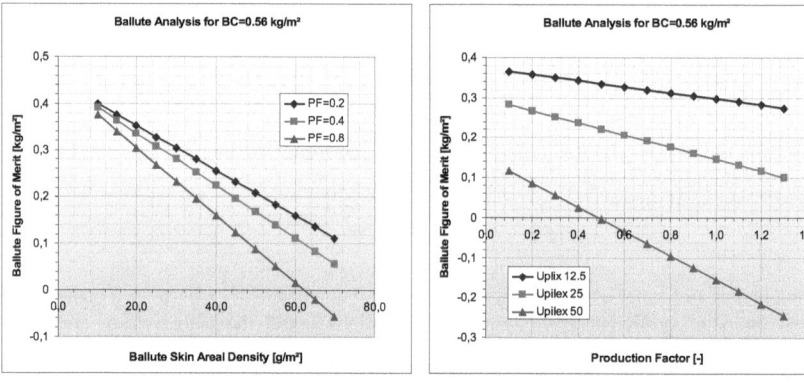

Fig. 7.34: Figure of Merit analysis for a ballute with a design ballistic coefficient of 0.56 kg/m² for various production factors (left) and areal densities (right) based on actually available films.

An analysis of this figure of merit for spherical ballutes, with a target ballistic coefficient of 0.56 kg/m², is given in Fig. 7.34. Recall that a ballute figure of merit of zero yields a ballute that meets the required ballistic coefficient without any Instrument Pod.

Equations (138) and (141) can thus be written in terms of the ballute figure of merit:

$$m_{ip} = B_{M,Sphere} \pi r_b^2 - \frac{p_{g,inf}}{R_g T_{g,inf}} \cdot \frac{4}{3} \pi r_b^3 \tag{145}$$

and for the simplification where $m_{ip}+m_H \gg m_{gas}$ is true as

$$r_b = \sqrt{\frac{m_{ip}}{B_{M,Sphere}\,\pi}} \qquad (146)$$

If $B_{M,Sphere}$ is large, the required ballute radius is small or the possible Instrument Pod mass is high. If $B_{M,Sphere}$ is small, the configuration becomes technically more challenging. If $B_{M,Sphere}$ is so small that the necessary ballute radius exceeds practical limits, the mission either has to be redesigned (raising the ballistic coefficient) or a higher performing skin film or better production methods have to be found. When we select thin film materials, we desire to know therefore how light the skin material has to be to achieve our mission. This can be found by solving (138) for the maximum allowable skin density.

$$\varrho_{A,max}' = \frac{3\beta c_d - \dfrac{3\,m_{ip}}{\pi\,r_{b,max}^2} - \dfrac{4\,P_{g,inf}\,r_{b,max}}{R_g\,T_{g,inf}}}{12(1+\lambda_{pf,min})} \qquad (147)$$

Here, $r_{b,max}$ is the maximum radius we want to allow for our ballute (size constraint) and $\lambda_{pf,min}$ is the best production factor we think we can realize. Since not all manufacturers provide the areal density of their products, here is a handy relation between it and the material density and film thickness:

$$\varrho_A' = \varrho_s \cdot \tau_s \qquad (148)$$

where ϱ_s is the material density of the skin film and τ_s the film thickness. Most manufacturers of films provide two of the three values.

Two things must be kept in mind when interpreting equation (147). First, $\varrho_{A,max}'$ is a theoretically determined allowable maximum, so available skin materials must be evaluated according to whether a strong enough skin can be made from them, which stays below that value. Secondly, skin stress was not evaluated so far! Let's recall that in all of the above, inflation gas pressure and temperature represent values at the time of inflation. They are carried along solely because we assumed that the inflation pressure is something that we want to define and need a relation that accounts for the resulting gas mass, as a function of the ballute size. So although one may feel tempted by (147) to obtain the skin density as a function of inflation pressure, it does not account for skin stress!

Therefore we are looking for the minimum possible areal skin density, that can withstand all loads. To find it, we first combine (135) and (134), to obtain the following equation for the skin density:

$$\varrho_s = \frac{m_s}{A_H\,\tau_s} = \frac{m_s}{4\,\pi\,r_b^2\,\tau_s} \qquad (149)$$

We then combine it with (90) and (92), to obtain an expression for the minimum skin mass possible:

$$m_{s,min} = \frac{2\pi\,\varrho_s\,P_{g,max}\,r_b^3\,\lambda_s}{\sigma_{s,max}} \qquad (150)$$

While this can be used to obtain the lightest possible ballute skin mass, it is not very practical, because it does not take areal densities into account which are available on the market. We therefore use equation (134) again and finally obtain the minimum areal density:

$$\varrho_{A,min}' = \frac{\varrho_s\, p_{g,max}\, r_b\, \lambda_s}{2\,\sigma_{s,max}}\,. \tag{151}$$

All that is left to do now is to find a material with a gauge such that

$$\varrho_{A,min}' < \varrho_{A,real}' < \varrho_{A,max}' \tag{152}$$

and the ballute design should be all set. Provided the material, if it can be found, can also be used for building a practical ballute (it needs to be seamed, folded, packed, stored, transported and deployed with available technical means and still maintain the properties for which it was selected). These problems are discussed in chapters 6, 8 and in particular 8.4.

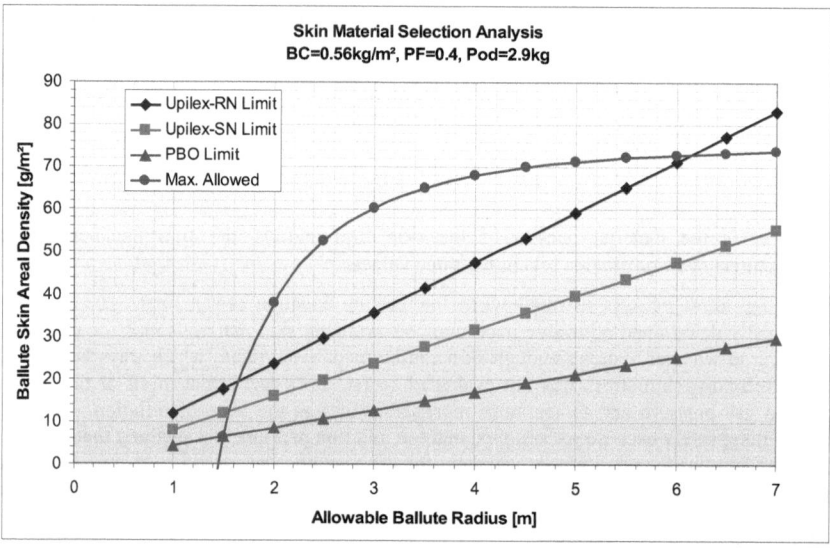

Fig. 7.35: Skin material selection analysis showing the maximum allowable areal density for a ballistic coefficient of 0.56 kg/m², a production factor of 0.4 and an instrument pod mass of 2.9kg, plotted over minimum areal densities for selected materials. Total safety factor is 3.2.

One of the above discussed analyses for MIRIAM is shown in Fig. 7.35 and one for ARCHIMEDES in Fig. 7.36. Both analyses assume a total safety factor of 3.24, which is a combined value from 1.8 each for λ_s on the tensile strength and the maximum pressure uncertainty. The selected material must be between the maximum allowable areal skin density curve and the minimum limit areal density for the selected material. Nominal interior pressures for the studied cases is 10hPa.

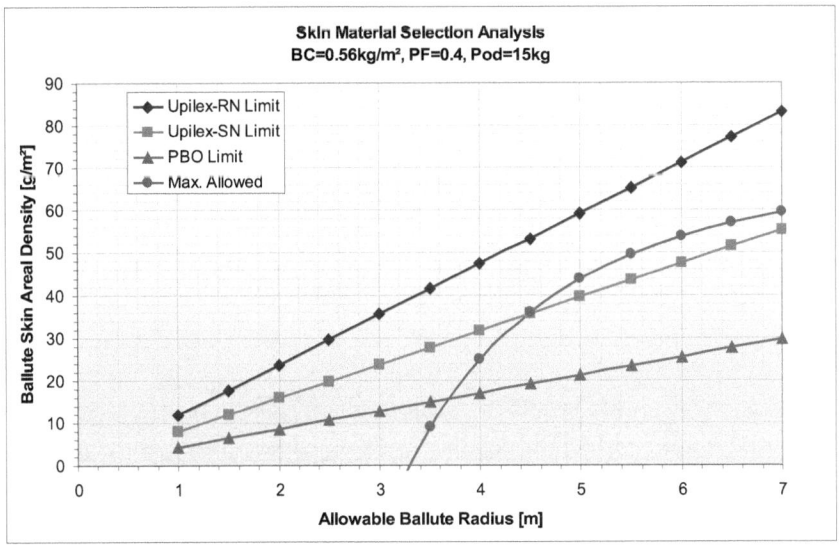

Fig. 7.36: Skin material selection analysis showing the maximum allowable areal density for ARCHIMEDES with a production factor of 0.4 and an instrument pod mass of 15kg plotted over minimum areal densities for selected materials. Total safety factor is 3.2.

Fig. 7.36 shows that Upilex-RN has insufficient strength for ARCHIMEDES, unless the production factor or the safety factors can be significantly lowered (Fig. 7.37) and that even a degraded PBO would the most desirable material choice.

When shopping for skin materials, one may find it useful to combine (147) and (151) with (152) to obtain the following expressions:

$$\frac{\sigma_{s,max}}{\varrho_s} > p_{g,inf}\frac{T_{g,max}}{T_{g,inf}} \cdot \frac{\left(1+\lambda_{pf,min}\right) r_{b,max}}{\beta c_d - \dfrac{m_{ip}}{\pi r_{b,max}^2} - \dfrac{4 p_{g,inf} r_{b,max}}{3 R_g T_{g,inf}}} \cdot \lambda_s \qquad (153)$$

for the material performance and, consequently,

$$\sigma_s = p_{g,inf}\frac{T_{g,max}}{T_{g,inf}} \cdot \frac{\left(1+\lambda_{pf,min}\right) r_{b,max}\, \varrho_s}{\beta c_d - \dfrac{m_{ip}}{\pi r_{b,max}^2} - \dfrac{4 p_{g,inf} r_{b,max}}{3 R_g T_{g,inf}}} \qquad (154)$$

for the skin stress. From either of these two it becomes very obvious why a light, yet strong skin material is so desirable.

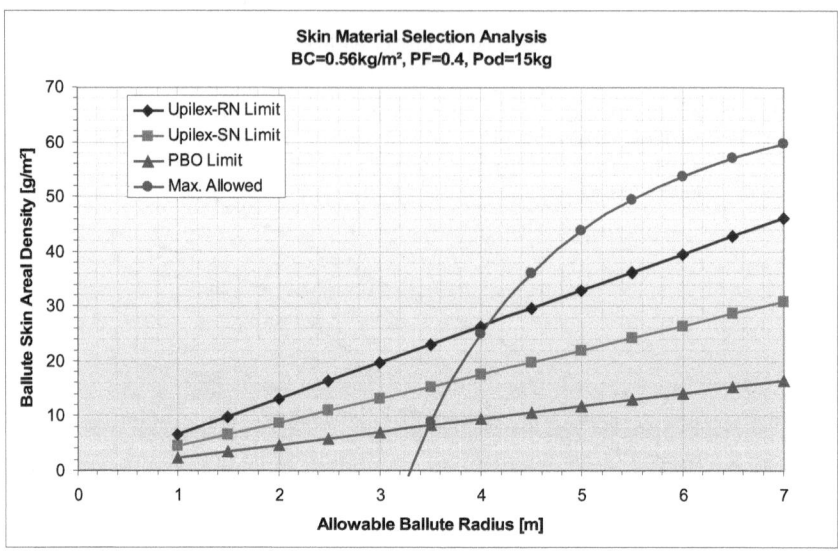

Fig. 7.37: Skin material selection analysis, showing the maximum allowable areal density for ARCHIMEDES, with a production factor of 0.4 and an instrument pod mass of 15kg, plotted over minimum areal densities for selected materials. Total safety factor is 1.8.

7.2.5   Sizing the Fittings

Like with ordinary balloons, the fitting is the area where the (usually hard) payload gondola (in our case the Instrument Pod) is attached to the (usually soft) skin and has to transfer the mechanical loads into the surrounding skin (Fig. 7.38). With our clamped pod design, the fitting is the rim of the instrument pod. This is shown in a principal cross sectional sketch in Fig. 7.39. As the ballute decelerates, it rotates to an attitude with the Instrument Pod facing forward in flight direction. The axial deceleration pulls the Instrument Pod forward, exerting stress on the ballute skin at the rim of the fitting (see Fig. 7.38 and also Fig. 7.16b). Assuming

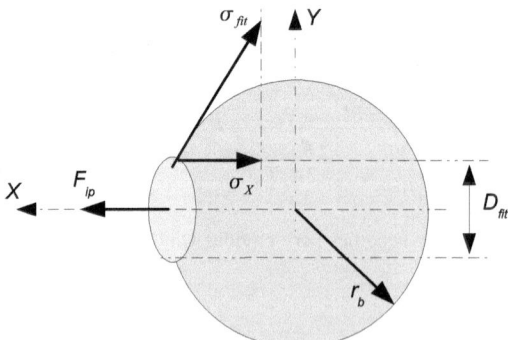

Fig. 7.38: Sketch of the fitting loads at the instrument pod.

a circular fitting, the equation that yields the skin stress σ_{fit} resulting from the fitting line load σ_{fit}' can be derived from Fig. 7.38, considering that the instrument pod load F_{ip} has to be distributed along the fitting perimeter:

$$\sigma_X' = \frac{F}{\pi D_{fit}} \quad \text{in} \quad \left[\frac{N}{m}\right] \quad \text{with} \quad \frac{\sigma_X'}{\sigma_{fit}'} = \frac{D_{fit}}{2r_b} \tag{155}$$

where D_{fit} is the fitting diameter. Fig. 7.39 shows the fitting in detail for a clamped instrument pod. Equations (155) can be rearranged to yield the skin line load at the fitting rim:

$$\sigma_{fit}' = \frac{F_{ip} \cdot 2r_b}{\pi D_{fit}^2} \tag{156}$$

To obtain the material stress, we have to distribute the perimeter line load across the available material thickness τ_s :

$$\sigma_{fit} = \frac{F_{ip} \cdot 2r_b}{\pi D_{fit}^2 \tau_s} \quad \text{in} \quad \left[\frac{N}{m^2}\right] \tag{157}$$

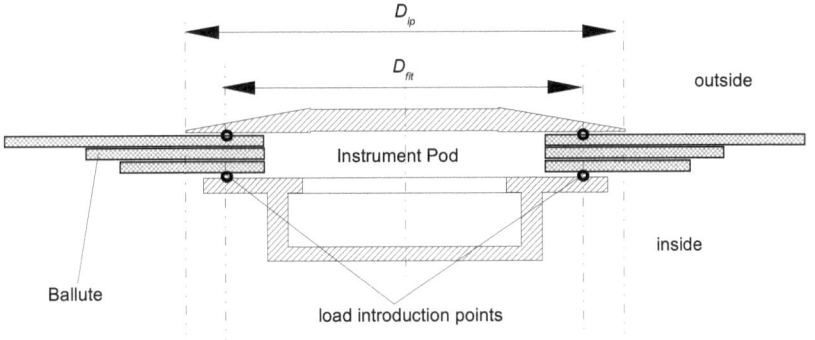

Fig. 7.39: Principal sketch of a cut through an instrument pod showing the load introduction point (fitting).

The fitting must be sized such that it can hold the Instrument Pod attached, when the highest force is acting on the Instrument Pod. This occurs at the point of maximum deceleration of the trajectory, so the maximum fitting stress is

$$\sigma_{fit,max} = \frac{m_{ip} \cdot 2r_b}{\pi D_{fit}^2 \tau_s} \cdot a_{sc,max} \tag{158}$$

The maximum deceleration is given with equation (75), so the maximum fitting stress may also be directly expressed in terms of entry parameters:

$$\sigma_{fit,max} = \frac{m_{ip} \cdot 2r_b}{\pi D_{fit}^2 \tau_s} \cdot \frac{v_E^2 \sin \gamma}{2H_0 \, e} \tag{159}$$

Results of the above are given for the ARCHIMEDES and MIRIAM ballute spacecraft in Fig. 7.40 without safety factors. Nominal fitting diameters for MIRIAM and ARCHIMEDES are 200mm and 600mm respectively.

The analysis shows that, if reasonable safety factors are applied and uncertainties in the real maximum deceleration force are taken into account (see 7.2.1.3), the pole area around the Instrument Pod fitting should be reinforced out to a diameter of at least 2.5 to 3 times the fitting diameter (see also chapter 7.3.3).

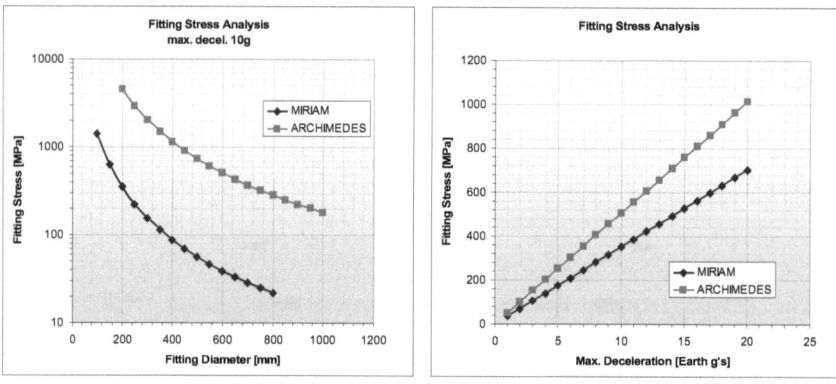

Fig. 7.40: Instrument Pod fitting stress analysis for MIRIAM and ARCHIMEDES.

7.3   Design Considerations

This chapter will briefly describe general design rules and considerations that have to be taken into account when designing a hypersonic drag ballute. They should be viewed as recommendations on possible ballute design, not necessarily ruling out other ideas.

All ballute spacecraft described herein were designed keeping these methods and rules in mind, unless otherwise noted.

7.3.1   General Layout Considerations

As the ballute enters the atmosphere, heat, aerodynamic and mechanical loads build up. These loads have a lasting and detrimental effect on polymer thin films, especially when they occur simultaneously. The ballute skin weakens and skin strain builds up more quickly. It is therefore necessary to mitigate any load peaks within the ballute skin. This can be done, by designing the ballute such that it can expand freely under pressure and reinforced areas, such as load introduction points and seams, should be made with a gradual, stepwise increase in thickness where possible.

To allow the ballute unhindered expansion, the windsock and hose should either hang freely or at least with a slack greater than the expected change in diameter (see Fig. 7.41).

Likewise, any instrumentation added to the ballute should be wired along the seams. These act as load tendons and have their own stiffness, depending on the design of the seam. Since even the seams expand more than a metal containing electric conductor, enough slack should be left in the wires to allow for some displacement.

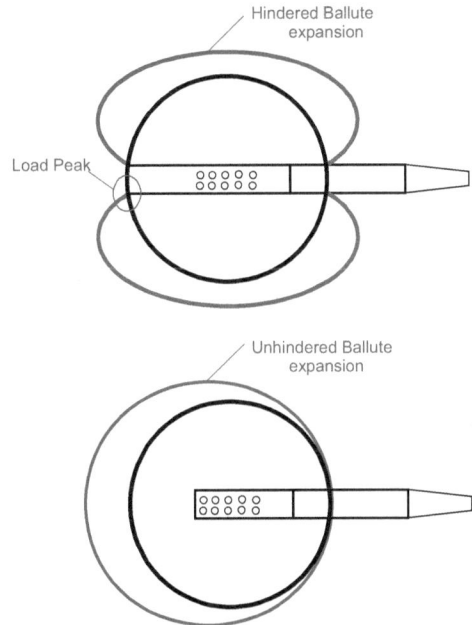

Fig. 7.41: Hindered and unhindered expansion of the ballute.

Of equal importance is that the ballute spacecraft's centre of mass will most likely lie somewhere in between the geometric centre and the Instrument Pod, causing the ballute to roll into an Instrument Pod forward attitude during hypersonic flight. This in turn leads to an increase in thermal load for the southern hemisphere. Any items protruding from the sphere, such as the inflation hose, should therefore be located at high northern latitudes, where aerodynamic forces are expected to be less violent and the aerothermodynamic environment is more forgiving.

7.3.2   Envelope Production Patterns and Segments

When we seek to fashion a closed three dimensional envelope from something two dimensional, we need to dissect the sphere into a suitable production pattern. The simplest would be a two segment cushion of the air bag design shown in Fig. 2.1, on page 6. Two skin sheets are held together by a perimeter seam. As discussed in chapter 2.2.4, such a crude approximation of a three dimensional body fashioned from two dimensional sheets causes a lot of skin strain, inherent stress and ultimately large wrinkles.

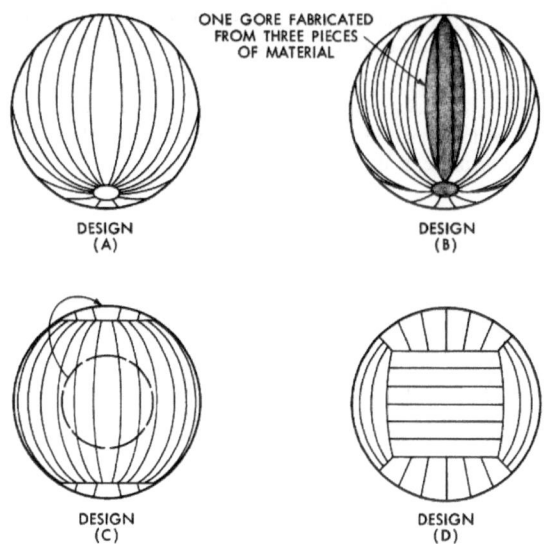

Fig. 7.42: Investigated balloon envelope designs of the Echo balloon programme [11]. (NASA)

If more segments are desired, a host of patterns become possible. Naturally, the stiffer the ballute skin material, the more segments are needed to approximate a sphere with tolerable stress levels. A toy balloon, made of highly elastic rubber, has just one bubble and no seams, whereas a ball made from paper will look angled, even with many segments.

Note that the number of segments is proportional to the number of seams needed, so an increasing number of segments also increases the production factor of the ballute and with it its weight. When choosing a production pattern, the possible number of segments has to be large enough to allow keeping skin strain below acceptable levels, but low enough for an acceptable production factor.

Last but not least, production and packaging patterns have to match to some degree, as the chosen patterns must allow an effective packaging method. Seams, especially double spliced sandwich seams (see Chapter 7.3.4), are much stiffer than the bare skin and can therefore not be folded as easily and as many times as a single skin layer. Production patterns where packaging folds run alongside of seams and cross them only a few times are clearly preferable.

During the development of the Echo balloons, a number of production patterns were tested [11], of which pattern design A from Fig. 7.42 (herein referred to as the "beach-ball" pattern) was selected. This pattern is favourable, because the seams direct mechanical loads in long meridians to the poles, which can be reinforced (see also chapter 7.3.2.3), whereas other patterns create load peaks at many points where seams end. For the prevailing study, patterns allowing very small skin elements were also studied, so as to investigate the applicability of PBO as a skin material, which may only be available in comparatively small sheets (see chapters 7.3.2.1 and 7.3.2.2).

7.3.2.1   The Football (Soccer) Pattern

This popular geometric figure is known to mathematicians as a truncated icosahedron and to the rest of the population as a football (or soccer ball). It is made from alternating pentagons and hexagons [80]. One of the many available paper models for such a pattern is shown in Fig. 7.43 (this pattern is public domain and no particular source could be found).

In this pattern every pentagon is surrounded by five hexagons.

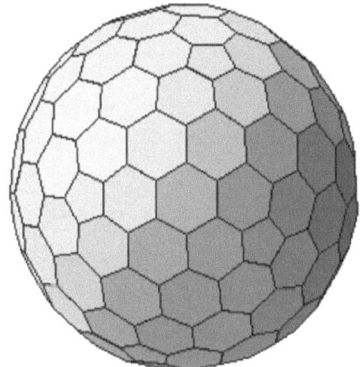

Fig. 7.43: The Football-Pattern (Public Domain).

Fig. 7.44: Polyhedron of a higher order made from hexagons and pentagons (Public Domain).

The advantage is that individual seams are short and that only junctions of three seams can occur. The entire grid of seams offers a relatively tight net, that can stop tears from propagating further along the skin and takes loads in all directions. It is also of considerable advantage that a polar cap is not necessary.

The disadvantage is that all elements are naturally flat and only a fixed number of segments is possible (20 hexagons and 12 pentagons) over the entire surface [80]. A pattern with criss-crossing seams is also more difficult to pack.

A variation on the theme would increase the number of hexagons, so that the spherical shape is better approximated as depicted in Fig. 7.44. Note that the number of pentagons (12) remains the same [80]. It is used for the dimples in a golf-ball and may therefore be referred to as the golf-ball pattern.

The advantage of this pattern is a better approximation of the sphere and better load distribution characteristics. Is is even conceivable to replace one element with an equally shaped instrument pod. In a hexagonal base geometry, such as MIRIAM and ARCHIMEDES, this shape would fit beautifully.

Note that due to the three dimensionally curved surface, not all hexagons have the same shape and angles (most prominently visible around the pentagons), making the manufacturing process difficult.

The large number of individual seams and the large number of triple junctions further complicate the manufacturing process and also lead to a rather large contribution of the seams to the production factor, driving up the overall system mass. This latter disadvantage can be an actual advantage if a material is found to be favourable that is not available in long rolls of thin film, but, like PBO, only comes in small sheets of limited sizes. In this case, a shape like the one in Fig. 7.44 can help to address this problem.

7.3.2.2 Geodesic Sphere

A geodesic sphere is generated by a network of great circles ("geodesics") across the surface, such that they intersect to form triangles [81]. Note that a similar method was used, in Chapter 5.4, to generate triangles of equal surface areas across the ballute for the thermal analysis model (see Fig. 7.45).

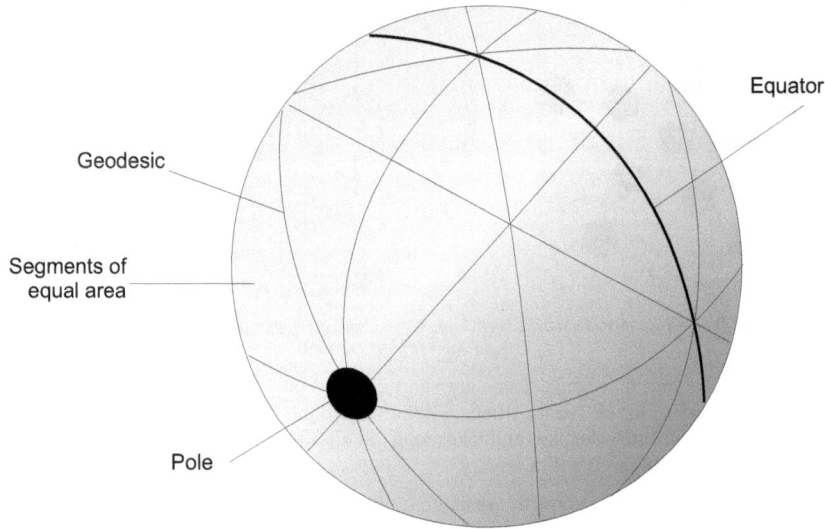

Fig. 7.45: A geodesic sphere with 48 elements.

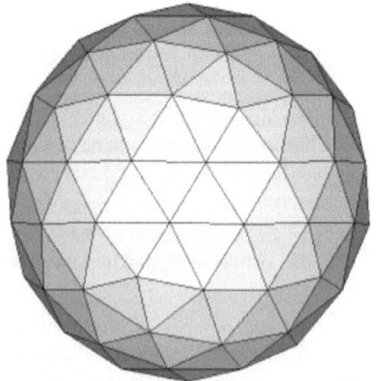

Fig. 7.46: Geodesic sphere with a reduced number of individual triangles (Public Domain).

A geodesic sphere can be obtained by starting with a simple Platonic shape, such as an octahedron or an icosahedron and further dissecting the triangular faces into more triangles. If done often enough, the resulting object approximates a spherical shell, made from flat face elements.

A great advantage is the optimal load distribution, as geodesic seams offer a globe-wide reinforcement grid of long and straight load lines.

Unfortunately, not all elements have exactly the same angles and edges of the same lengths. This makes production cumbersome and prone to errors. The problem can be eased somewhat by accepting that vertices do not necessarily have to line up into perfect geodesics. This leads to a design pictured in Fig. 7.46. While offering fewer individual shapes, care must still be taken upon assembly. Note that some triangles form pentagons and others form hexagons, showing differences in angles! Another notable disadvantage is the multitude of seams and the many intersections with five or six seams meeting in big bulges all over the surface.

However, this design may still be considered as a viable option, if only small film elements can be obtained.

7.3.2.3   The Beach Ball (Orange Peel) Pattern

This production pattern is used for most spherical envelopes, including almost all balloons, parachute canopies, inflatable party domes and beach balls. It was also chosen for the Echo balloons [11]. The pattern is akin to an orange which is peeled by running a knife through its skin from tip to navel. It is also akin to the geodesic pattern, as it consists of seams in great circles, but only running along meridians and the equator. This leads to a huge build-up of seams at the poles, so it requires polar cap segments. On the south pole, the cap is replaced by the instrument pod. As an example, the 36-segment 10-m ARCHIMEDES ballute is shown in Fig. 7.47.

Fig. 7.47: Artist's impression of the ARCHIMEDES 10-m ballute on the surface of Mars.

This pattern is especially suitable for a skin material, which is available in very long sheets. Only the curvature along latitudes needs to be approximated by a polygon, as the film is allowed to bend naturally along meridians.

The advantage is clearly that when entering the atmosphere with a pod-forward attitude, the flow field "sees" a perfectly round object (safe for manufacturing imperfections and shape deviations). From a structural point of view, the advantage is that the pod load is directly distributed to the stronger seams, which act as load tendons, if designed accordingly. They also direct mechanical loads towards the poles, which necessitates polar cap reinforcements.

In the design department, the beach-ball pattern is the only one that lets us freely choose the number of elements, all of which are identical and production is comparatively easy. Last but not least, longitudinal seams make the ballute more easy to fold up into a high efficiency package. It should also be mentioned that, due to its wide use, there is a large and common body of knowledge.

Because of the above considerations, the production pattern for all test ballutes has been the beach ball pattern. The only major disadvantage is that longitudinal tears in the skin can theoretically propagate along the entire length of a hemisphere (until being stopped by a polar cap or the equator). Without further proof, a segment contour can be calculated with the following formula:

$$y = \frac{\pi \cdot d_B \cdot \sin\left(\frac{2x}{d_B}\right)}{n} \tag{160}$$

where d_B is the diameter of the ballute and n is the number of meridian segments around the equator. The latter number has to be multiplied by two to obtain the total number of segments, if the ballute has an equatorial seam. So, for a 32-segment balloon with an equator seam, 64 individual segments have to be made. The two sets are likely to be different, as the north polar segment has to have a different clearance radius for the polar cap. The contour is plotted for a 4-m diameter ballute with 32 meridian segments on two hemispheres in Fig. 7.48.

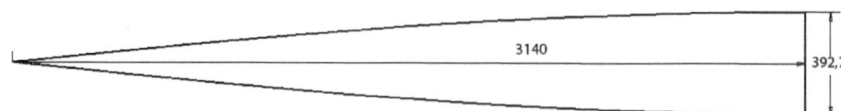

Fig. 7.48: Segment contour of a 32-segment 4-m ballute as used on the MIRIAM space flight test.

To leave the north polar cap in one piece, the inflation hose cannot pass into the ballute right at the pole. On the MIRIAM ballute, which used that pattern, it was installed at about 85° North. That way a longitudinal seam could be used as the feed-through point into the ballute.

7.3.2.4   The Necessary Number of Segments

Whatever the pattern, the necessary number of segments has to be determined to obtain the optimal compromise between inherent skin strain, due to stretching a polyhedron into a sphere (by applying pressure) and the number of seams. For a ballute made from a polyimide or similarly stiff material, we choose to be consistent with the definition of the $R_{p0.2}$ stress limit, which means a plastic deformation of less than 0.2% is considered small enough to represent an acceptable minimum. To go along with that definition, we find a 0.2% stretch to be an equally acceptable limit for our skin pre-loading.

We find the pre-loading strain by considering what strain pre-loading would be necessary to stretch the polyhedron or orange peel polygon, into a perfect sphere. We do that by comparing the circumference of the segment polyhedron to the equivalent design sphere.

When designing a soccer ball, we consult literature on archimedean solids [81][82], to find the radius of the outer bounding sphere of a truncated icosahedron, which is

$$r_{outer} = r_b = \frac{s}{4} \cdot \sqrt{58 + 18 \cdot \sqrt{5}} \tag{161}$$

where s is the side length of one edge. The assumption that the design ballute radius should be that of the outer bounding sphere of the archimedean solid is justified, because under interior pressure, any object tries to become round, hindered only by skin tension, and so the segments will bulge outward. Because the actual circumference of the truncated icosahedron depends on the path we follow around, we set the radius of the outer bounding sphere into relation with the inner bounding sphere and assume that the average lies in the middle somewhere.

The radius of the inner bounding sphere is

$$r_{inner} = \frac{s}{4} \cdot \sqrt{3 + 3 \cdot \sqrt{5}} \tag{162}$$

and the relation between the inner and outer bounding sphere is

$$\frac{r_{inner}}{r_{outer}} \approx 0.91 \tag{163}$$

With the difference therefore being around 9% and the real circumference being roughly in the middle somewhere, the skin will have to stretch up to 4.5% to become spherical. From the stress-strain diagram we can now see how much pre-loading of the skin is necessary to achieve a spherical ballute. While this way of predicting the tension pre-loaded into the ballute through the strain of its shape is not very accurate, it gives us the important information that the soccer-ball-ballute is not a very suitable solution, if we seek to keep skin strain at bay.

For the beach-ball pattern, we take a look at the sphere and observe that the shape departs from a perfect sphere most notably at the equator. If we dissect the sphere here and look at it head on, we see a regular polygon with n sides, of which the perimeter length is obviously

$$l_{Perim} = n \cdot s \tag{164}$$

where n is the number of segments and s is the side length of one segment. The length of a segment side is

$$s = r_b \cdot 2 \cdot \sin\left(\frac{\pi}{n}\right) \tag{165}$$

To find the pre-loaded skin strain necessary to make the polygon into a circular equator, we set into relation the polygon perimeter length and the equatorial circumference of the ballute:

$$\frac{l_{Prim}}{l_{equat}} = \frac{n \cdot r_b \cdot 2 \, \sin\left(\dfrac{\pi}{n}\right)}{2 \, \pi \, r_b} = \frac{n \cdot \sin\left(\dfrac{\pi}{n}\right)}{\pi} \tag{166}$$

and find the percentage by which the material has to stretch through

$$\frac{\Delta l}{l} = \left(1 - \frac{l_{Perim}}{l_{equat}}\right) \tag{167}$$

Since the meridian seams take away some of the segment base lengths and are not as elastic as the base material, due to the added thickness of the tape, the seams leave a shorter portion of the material to do the stretch. So the above equations give a theoretical minimum.

If we assume a double spliced seam (see Chapter 7.3.4) on either side of the segment and further assume the taped seam to contribute nothing to the strain, each segment looses one tape width for the stretch. We further assume that the seam width is small as compared to the radius so that its equivalent circle section has the same length. With the number of vertices being equal to the number of polygon segments and therefore having the same number of segments and seams, we can rewrite equation (167) to include the tape width in the analysis:

$$\frac{\Delta l}{l} = \left(1 - \frac{l_{Perim} - n \cdot l_{TapeWidth}}{l_{equat} - n \cdot l_{TapeWidth}}\right) \tag{168}$$

If the tape width is also small as compared to the segment base side length (in case of MIRIAM the tape width was 25mm), the addition in strain is around 0.04% or less for relevant ballute designs and can therefore be neglected. Results of the above equations are shown in Fig. 7.49.

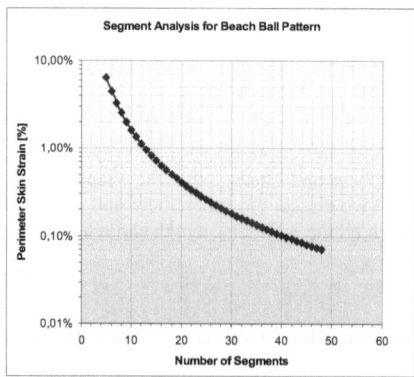
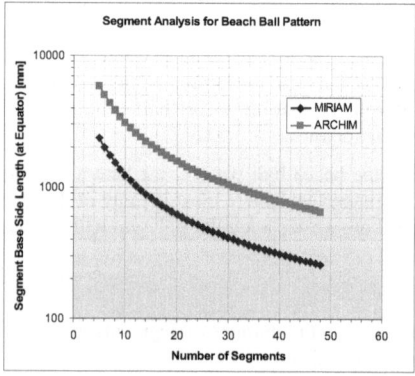

Fig. 7.49: Theoretical minimum strain and segment base lengths for the beach ball production pattern.

To remain below the 0.2% strain limit, we find that the number of segments has to be at least larger than 30. To remain sufficiently clear of it, we need at least 32 and more if we can afford that. Note that more segments lead to more seams and shorter equatorial segment base lengths.

7.3.3   Fittings and Reinforcements

The reinforcements around the Instrument Pod fitting not only strengthen the skin material locally, but also mitigate load peaks around the load introduction area and thus lead to a smoother stress distribution across the skin area. If more than one reinforcement blanket is

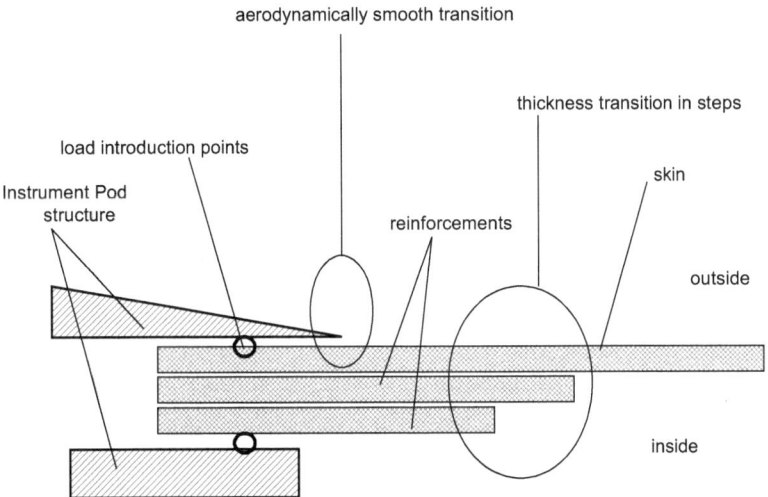

Fig. 7.50: Sketch of a principal cross-section through the rim of an Instrument Pod fitting. The pod extends to the left, the skin continues to the right and the skin is viewed head on.

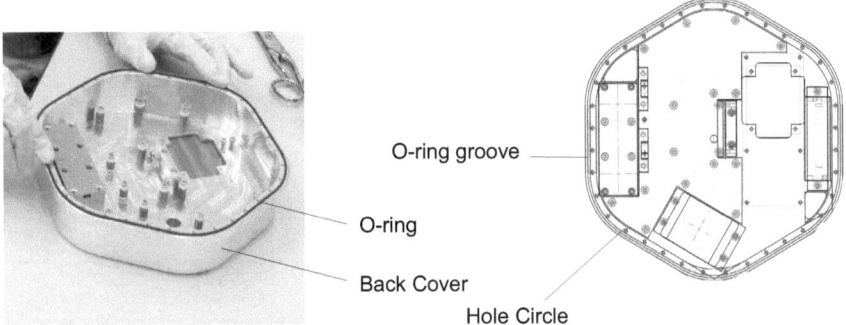

Fig. 7.51: MIRIAM's Instrument Pod Fitting

bonded to the ballute skin, shear lag theory advises us that abrupt transitions in thickness should be avoided. Instead the thickness should be reduced in steps (see chapter 7.3.4).

A principle sketch showing the cross-section of an Instrument Pod fitting is given in Fig. 7.50 (see also Fig. 7.39). When introducing loads into the ballute, care should be taken that the fitting transfers these loads uniformly or load peaks will consequently be created. At the same time, the fitting perimeter should be made gas tight, to keep the hot hypersonic boundary layer out and the ballute inflation gas in.

To do that, the MIRIAM Instrument Pod had an inner hole circle surrounded by an outer O-ring groove. The O-ring kept the ballute gas-tight and at the same time created a seamless load introduction line (see Fig. 7.51).

7.3.4 Seams

The seams have to bond the ballute segments together and have to meet or exceed the performance of the actual skin material. Seams can be either made with adhesive tapes, stitching, welding or a combination thereof. Because stitching creates holes from which cracks can propagate and, therefore, weaken the thin film material required for the application at hand, stitched seams are not further investigated here. They might, however, be useful in other ballute applications.

7.3.4.1   Seam Geometry

This chapter will deal with the general geometry of the seams and how to calculate seam parameters, as well as discussing the impact of seam design on the manufacturing process.

The simplest of seams, which is also easy to manufacture, would be a perimeter seam, as shown in Fig. 7.52. It is manufactured by placing two segments on top of each other and then seaming along the perimeter. The method is used by tailors to put clothes together which are manufactured inside-out, so that the later piece of clothing hides the stitched seam on the inside. The first test balloons made from co-extruded three-layer polyethylene were made that way (see Fig. 7.53), because placing two segments on top of each other makes a flat seam which can easily be welded with a hand welding device (see also Fig. 7.63 in chapter 7.3.4.3).

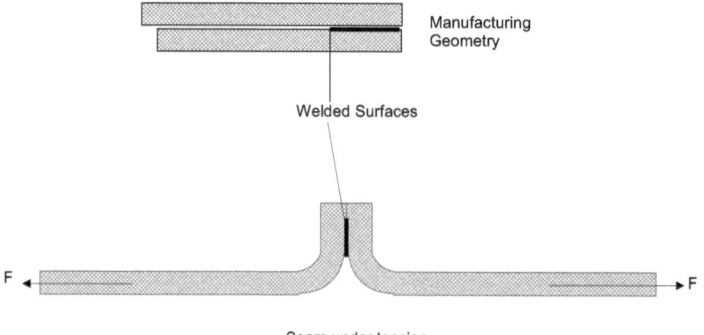

Fig. 7.52: Geometry of a simple perimeter seam.

For an actual ballute that has to withstand high loads and has to have an aerodynamically smooth surface this technique is not practical. On a spherical ballute, the last seam has to be manufactured outside out, leaving at least one seam to stand out as an aerodynamic disturbance. Additionally, the tension is causing the seam to stand up and peal. Such a seam can tolerate much less stress than seams subjected only to shear. Lap-joints are, therefore, a much better solution for the task at hand.

Fig. 7.53: Welded seams on a polyethylene balloon and inflation hose, manufactured by placing two segments on top of each other and welding along the perimeter. Note the seam standing out.

A simple lap-joint seam (made by overlapping segments) is shown in Fig. 7.54. Unfortunately, misaligned force vectors cause the seam to rotate, adding a peeling stress to the shear stress between the bonded surfaces. This effect significantly weakens the seam and causes scalping.

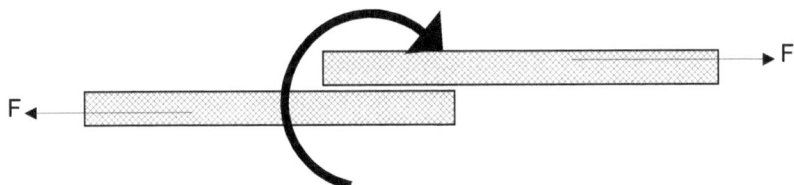

Fig. 7.54: Overlapping segment lap-joint.

Classic shear-lag theory (given in standard text books such as [83] and [84]) tells us that a taped seam transmitting a pulling force through shear stress, does not have an equal load distribution. The bulk of the load is taken by the area close to the edge of the tape (see Fig. 7.55). Therefore, simply increasing the surface to make the seam stronger, won't work beyond a certain limit width. A good solution to all of these problems is a double spliced seam with two tapes, as shown in Fig. 7.55. It aligns the force vectors and thus prevents rotation of the joint and doubles the surface which transfers the load through shear.

The double splice, however, creates a high stress peak at the point where the overall skin thickness changes from one layer to three layers. Such areas are prone to material failure.

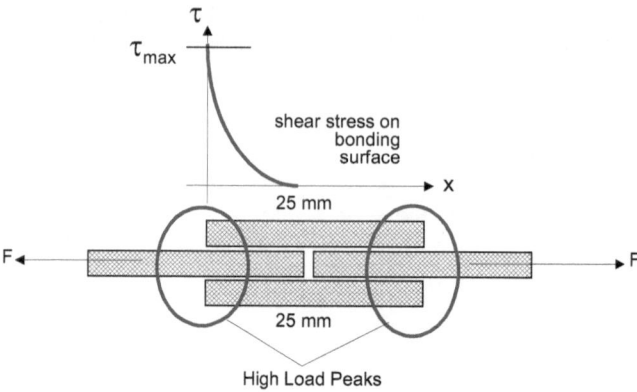

Fig. 7.55: Double splice seam with two equally wide tapes.

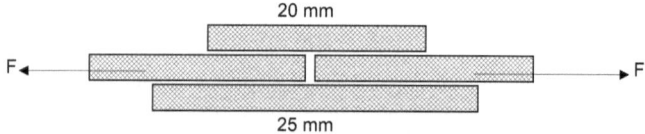

Fig. 7.56: Double spliced seam with two differently wide tapes.

An even better solution is therefore a double spliced seam, with a narrow and a wide tape, as shown in Fig. 7.56. This increases overall skin thickness from one to three in steps, therefore minimizing load peaks (note that the same technique of load peak mitigation has been described in chapter 7.3.3, for skin reinforcement layers).

7.3.4.2   Seams with Adhesive Tapes

Two splice lap joints of the design shown in Fig. 7.55 and Fig. 7.56 can be manufactured using adhesive tapes. Adhesive tapes have the advantage that they are ready to use and can be applied with comparatively little effort (in the simplest case by hand). The disadvantage is the limited thermal stability and durability of most adhesives, anyway of those which are readily available on the market. In addition to this, most adhesive tapes do not conform to outgassing requirements of space applications and seams which are not properly taped up may stick to unwanted skin segments during packing.

A seam with an adhesive layer can be calculated based on shear theory. Since the adhesive layer bonding tape substrate and segment transfers the skin tension mainly through shear stress, the shear load in the layer has to be calculated and compared against the adhesive's shear limit. The following differential equation for laminates of two layers bonded by a glue layer can be used to do this [84]:

$$\frac{\partial^2 \tau_{gl}}{\partial x^2} \cdot \tau_{gl} \frac{G_{gl}}{t_{gl}} \left(\frac{1}{E_1 t_1} + \frac{1}{E_2 t_2} \right) = 0 \tag{169}$$

where τ is the shear stress , E is Young's modulus for the skin and tape substrate in [Mpa], G is the shear modulus in [Mpa] and t is the layer thickness, with subscripts gl indicating the glue and 1 and 2 the respective layer (skin and tape substrate). Assuming that skin material and glue remain in the linear-elastic realm and that the skin stress is transmitted only through shear in the glue, a closed analytical solution can be found. We can apply this to our problem, by dividing the skin into 3 two-layer sections (Fig. 7.57), by treating the third layer, which is not contributing to the shear stress, as an addition in layer thickness. For each section, we then get the following analytical solution:

$$\tau_{gl} = \frac{1}{\alpha} \cdot \frac{G_{gl}}{t_{gl}} \cdot \frac{\sigma_1}{E_2} \cdot \frac{\sinh(\alpha l_{sec} - \alpha x)}{\cosh(\alpha l_{sec})} \quad \text{in [MPa]} \tag{170}$$

with

$$\alpha = \sqrt{\frac{G_{gl}}{t_{gl}} \left(\frac{1}{E_1 t_1} + \frac{1}{E_2 t_2} \right)} \tag{171}$$

where l is the section length, t is the layer thickness and x is the variable starting at the beginning of the section.

From this set of equations, it becomes obvious that a gooey adhesive layer with a low shear modulus and a thin tape substrate mitigate load peaks, but also weaken the seam. So the optimal taped seam has enough tape substrate thickness to match the stress of the skin over the gap where the two skin layers are not connected. In other words, if skin and substrate are made from the same polymer with roughly the same mechanical properties, the double splice should have half the substrate thickness of the skin, because over the gap two tape layers carry the stress, which is otherwise carried by one skin layer. A thicker tape would add material, but make the seam unnecessarily stiff at the expense of added shear load in the glue. A thinner tape would ease the shear stress in the glue layer even further, but material thickness would be insufficient to negotiate the load over the gap.

Results for a 10-m UPILEX-25RN ballute assuming a 10 hPa interior overpressure and seams made from UPILEX-12.5RN adhesive tapes are given in Fig. 7.57. Calculations are further based on the assumption of a glue layer thickness of 50μm and a shear modulus of 115MPa for a standard adhesive layer and 20MPa for a much weaker (more "gooey") layer for comparison.

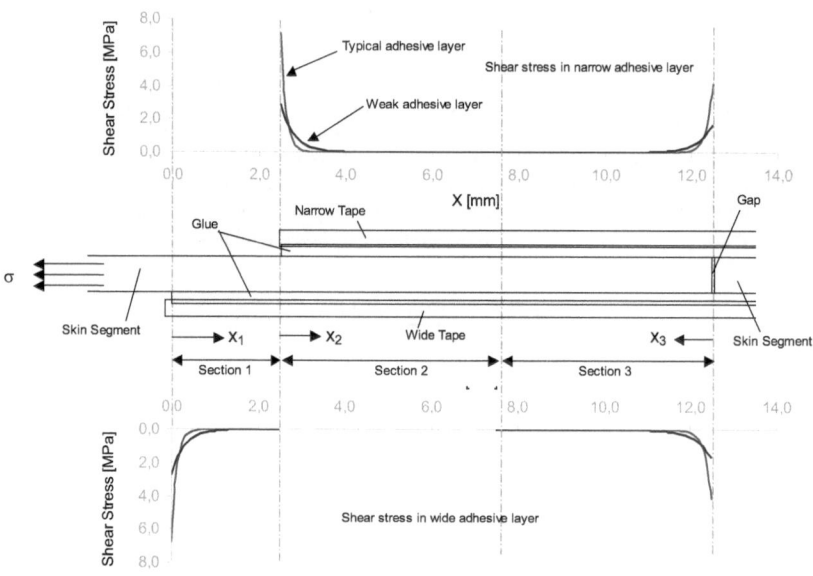

Fig. 7.57: Analytical shear stress analysis of a two-splice lap joint seam, based on a 10-m UPILEX-25RN ballute, with an interior overpressure of 10hPa and 12.5μm UPILEX-RN adhesive tapes.

For comparison, Fig. 7.58 shows the same seam with a tape substrate thickness of 25μm. Load peaks are around twenty percent higher as compared to the design shown in Fig. 7.57, with a tape substrate half as thick. Note that load peaks over the segment gaps are considerably higher as well.

If the same seam is built with equally wide tapes, both layers will have high load peaks, as the stiffness transitions abruptly from one to three layers (see Fig. 7.59).

Results are typical silicon-based glue values for room temperature, so the graphs represent typical curves. For material property values of a quality befitting reliable seam design the precise thickness, the shear modulus and the maximum allowable shear stress over the mission temperature range need to be known.

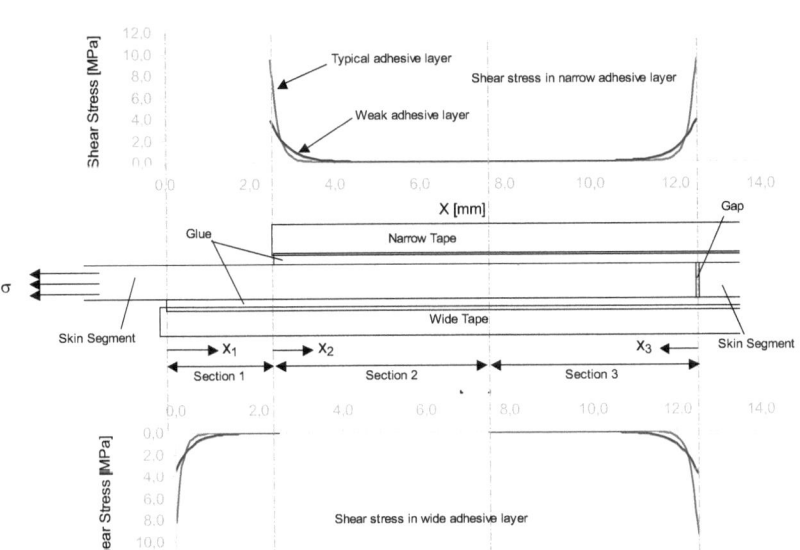

Fig. 7.58: *Analytical shear stress analysis of a two-splice lap joint seam based on a 10-m UPILEX-25RN ballute with an interior pressure of 10hPa and 25μm UPILEX-RN adhesive tapes.*

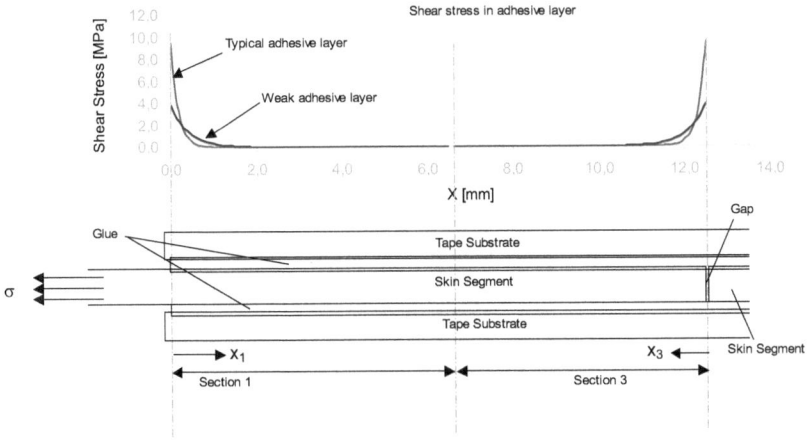

Fig. 7.59: *Analytical shear stress analysis of a two-splice lap joint seam, based on a 10-m UPILEX-25RN ballute, with an interior pressure of 10hPa and two 25mm wide 25μm UPILEX-RN adhesive tapes. Only one side is shown, as the shear stress in the other side's adhesive layer must be the same, because of the symmetry.*

But the graphs do show the typical load peaks and also that only the first two millimetres of the tape width really contribute to the transfer of stress, even in case of a weak glue and that the rest is more or less redundant. In case the narrow tape width is 20mm, the "safety factor" is 10. This "safety factor" redundancy is needed to account for a reduction in available load carrying surface area, due to manufacturing imperfections, segment dislodging under stress and various misalignments that can be minimized, but are inevitable during real world production. Possible imperfections and flaws are shown in Fig. 7.60 and Fig. 7.61.

The graphs also show that it is a good idea to reduce the tape width on one side of the seam, combined with reducing the substrate thickness to half the thickness needed to carry the same tension as the skin segment (because there are two tape layers). This seam design not only reduces production weight, but also helps lowering load peaks, making the seam more durable.

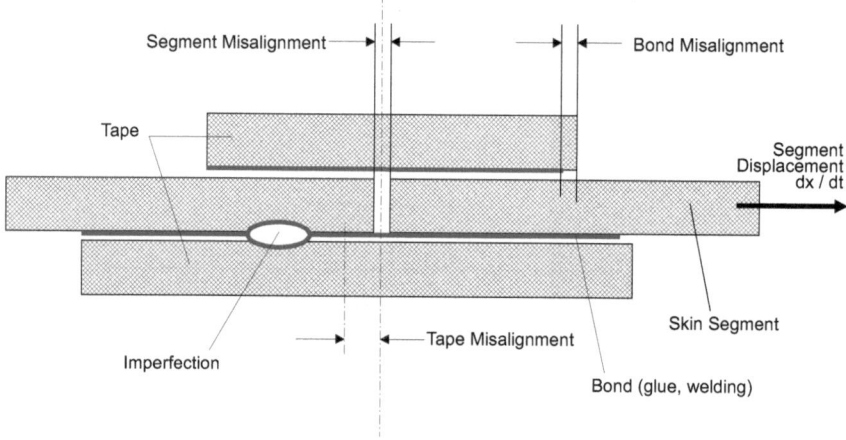

Fig. 7.60: Imperfections and production flaws in a real-world double spliced seam.

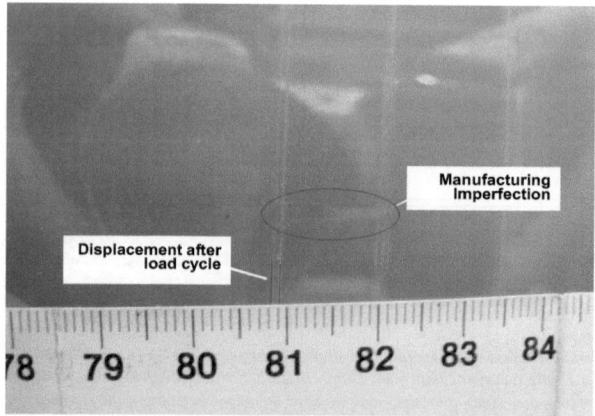

Fig. 7.61: Typical segment displacement and a flaw after a load cycle on a seam test specimen. Scale is in centimetres [64].

The in-house research laboratory of the commercial company Lohmann Tapes of Neuwied, Germany manufactured a custom made 25-mm adhesive tape based on an UPILEX substrate, trying to make the adhesive layer as tough and as heat resistant as possible. Laboratory tests with sample segments of two-spliced seams with that tape were made, subjecting the seam to biaxial dynamic load with a biaxial static offset and heating. The test showed that the seam strength exceeded that of the UPILEX-RN thin film under room temperature, but failed under temperatures around 350K [64]. The resulting damage is shown in Fig. 7.62. In order to fulfil the ARCHIMEDES temperature and mechanical load requirements, adhesive tapes alone must therefore be declared insufficient.

Unfortunately, the silicon-based adhesive did also not conform to outgassing requirements for space applications. This might not be a problem if it can be proven that the amount of adhesive exposed to space (at the fringe of any seam or reinforcement) is lower than a certain minimum amount.

As far as the 4-m diameter ballute for the space fight test MIRIAM is concerned, an investigation by IABG of Ottobrunn, where the flight model ballute was tested inside their 6-m space simulation chamber, revealed that the amount of adhesive exposed to the vacuum environment is small enough to be acceptable. No values were given, when permission was granted.

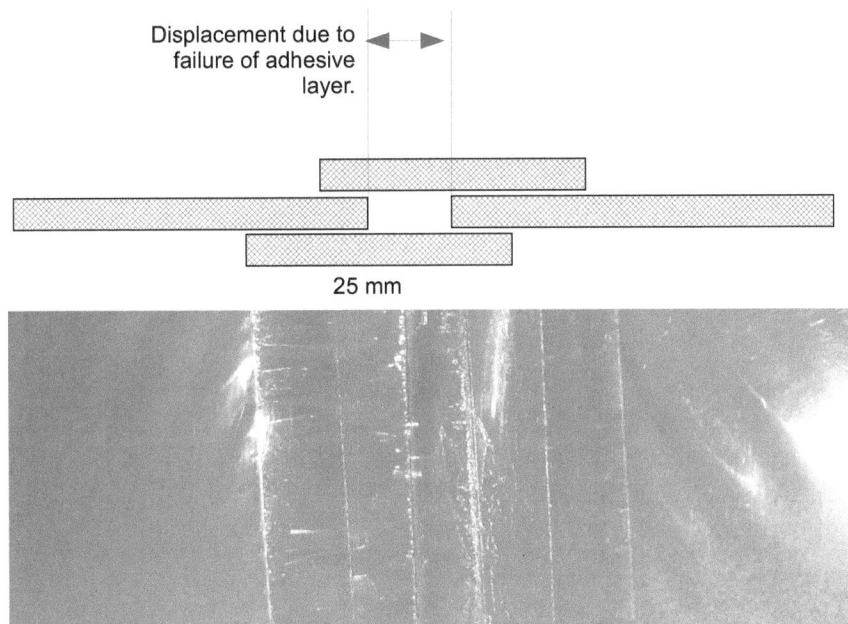

Fig. 7.62: Damaged seam due to a failure of the adhesive layer at 350K during a biaxial load test with a skin tension of 100MPa.

7.3.4.3   Heat Bonded (Welded) Seams

One possible solution to the challenge of manufacturing durable seams is heat bonding (or welding). It works by simultaneously applying pressure and heat, both carefully tailored to the polymer's properties. In contrast to adhesive layers and classic welding, heat bonds soften the polymer's surface, causing polymers of either surface to mingle with the other under the applied pressure, thereby interlocking and forming a durable bond after the joint cools off. Naturally, this works best with two identical thermoplastic polymers that easily melt and solidify. It may also work with a combination where one film is made from a thermoplastic polymer and the other from a thermoset plastic. In this case the melting temperature of the thermoplastic must be sufficiently lower than the melting point of the thermoset, to prevent irreversibly damaging the latter in the welding process.

Heat bonding has the advantage that no adhesive layer is needed, reducing the seam weight and bringing the production factor down. It has the disadvantage that no adhesive layer is available, making considerable mechanical effort necessary to produce a precision seam of constant and reliable quality. The main advantage, however, is that the shear strength of the resulting bond is the main contributor to the seam strength. It can be as high as the weakest member of the heat bonded compound.

Any of the seam types discussed in this chapter can also be welded. Fig. 7.63 shows the welding of a perimeter seam on the inflation hose of a polyethylene test balloon with a hand-held rolling welder. Fig. 7.64 shows how lap joints can be manufactured, either by bonding the tape to the segments, by bonding the segments directly or by using a special bonding polymer (heat bond layer) in place of the adhesive layer discussed in chapter7.3.4.2 .

Welded two splice sandwich seams (see Fig. 7.65) were made and tested with UPILEX-25RN segments and tapes, using 6µm Norton® PFA thin film (a variant of Teflon®) manufactured by Saint Gobain Plastics of Worcester, Massachusetts [85] as a heat bonding layer. Welding was done at temperatures between 570K and 620K, with two heated hand-welding irons, by placing the seam in-between the two and applying pressure [86]. Subsequent load tests at 470K with dynamic load cycles of 30Hz and 10MPa amplitude peak to peak showed no seam failures over the entire expected mission time. Neither did static load tests of that seam type.

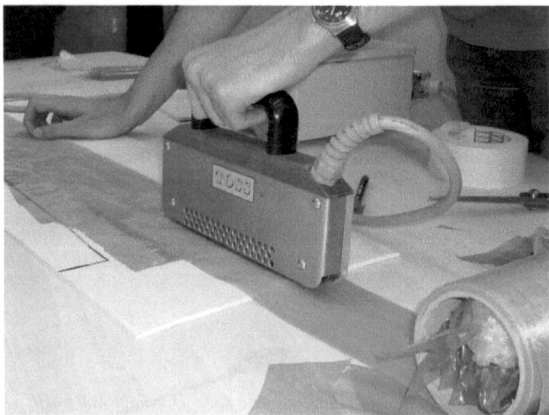

Fig. 7.63: Hand welding of two polyethylene layers.

At a dynamic amplitude of 30MPa, the bare UPILEX around the seams which were welded at 620 K failed. This suggests that the heat during the welding process weakens the compound layers and that the welding temperature should not exceed a limit beyond which the original material properties are adversely and irreversibly affected. This limit is a matter of experiment evaluation and, in the case of the UPILEX-RN / PFA compound, needs to be verified in a further test series, involving a more automated welding process that can provide better control over parameters.

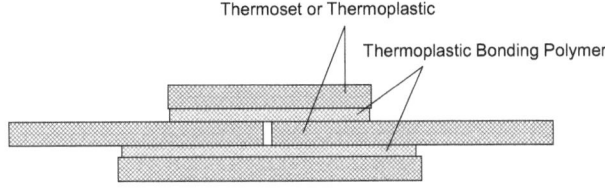

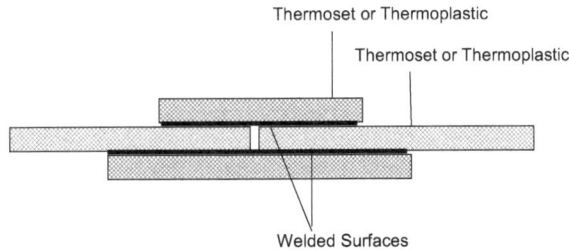

Fig. 7.64: Welded lap joint with a heat bonding layer (top) and directly welded layers (bottom).

The performance of the UPILEX/PFA compound is anyway far superior to that of the best adhesive layer available. The seam is thinner, lighter and complies with space flight outgassing requirements. Unfortunately, the manufacturing process is rather complicated, as necessary temperatures are beyond the capabilities of commonly available hand welders and welding machines for thin films. In addition to this, all elements of the seam (two tapes, two segments and two bonding films) have to be carefully aligned, before applying pressure and temperature, to ensure no film is standing off, weakening the seam and compromising the smoothness of the ballute's surface. A test run of this complicated process was made by producing a long continuous seam at the company Ballonbau Wörner of Augsburg, Germany with their special custom-built high temperature continuous welding machine. The seam came out very well, far exceeding the hand welded samples in quality and performance. Manufacturing a full ballute with this technique would require further modification of the machine's tape feeding system, so it can process the two additional bond layer films. In the seam test, this was done by hand, but did not come out very precise. Building such a mechanism will be difficult and not supported by Wörner. Rebuilding the machine altogether will be impossible, because to this date Ballonbau Wörner holds technical details pertaining to the machine and its principle parameters confidential. No pictures were allowed to be taken. An agreement could be reached that upon specifying temperature, pressure and dwell time, Wörner saw himself able to manufacture a ballute to our specifications, if no technical modification of the welding machine's mechanisms are necessary.

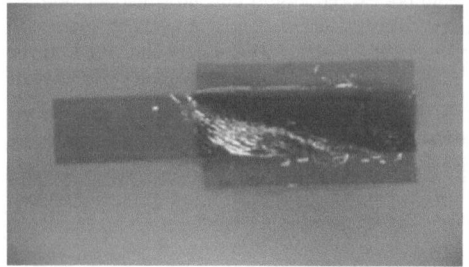

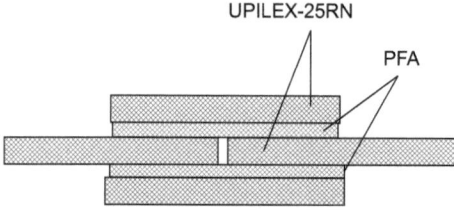

Fig. 7.65: Welded lap joint with UPILEX-25RN segments and tapes heat-bonded by a 6μm PFA thin film..

Another possibility would be to use a tape with the welding layer already bonded (if only weakly) to the tape substrate. Tests by Lohmann Tapes to coat the UPILEX with PFA before cutting the tapes were unsuccessful. Saint Gobain Plastics already offers a heat-sealable composite of their own polyimide variant with FEP (fluorinated ethylene propylene) coating [87]. The composite is marketed under the trade name of FluoroWrap™. Unfortunately, FEP does not deliver the performance required for the ballute [88] and neither does their polyimide variant Norton® TH [89]. Saint Gobain Plastics developed a new composite based on FluoroWrap™ that features PFA in place of FEP [90]. Coating of UPILEX with their PFA would be technically possible for them, but since UBE of Japan is a direct competitor of Saint Gobain, only their own polyimide will be available with PFA coating. Naturally, the coating technique is a patent of theirs and as such confidential.

So for the time being, either a high temperature plastic welding machine with the necessary continuous feeding system has to be constructed or a method has to be found, how to bond tape and welding layer. Or a different seam welding technique has to be developed.

During material tests with PBO [75], it was accidentally discovered that it can be welded to UPILEX to form a durable bond without the necessity of a bonding layer in between (see Fig. 7.66). This design has the advantage of being even lighter and thinner and the very durable PBO tapes add a light yet rigid load cage to the ballute that has a considerably better performance than the other seaming types.

A major test series, like the one done for welding the PI/PFA compound and the adhesive tape joints, was not completed before the conclusion of this thesis, but the preliminary data looks very encouraging. Manufacturing a lap joint, as shown in Fig. 7.66, could be done by Ballonbau Wörner without much modification to their specialized equipment, yet the resulting seam would be far superior to any of the other seams discussed above. Alas, PBO cannot be manufactured in lengths necessary to produce continuous seams spanning a quarter globe of ten metres (or even a four metre globe for that matter). While technically possible, the required equipment to do so does not exist, so considerable investments are necessary to obtain long PBO tapes.

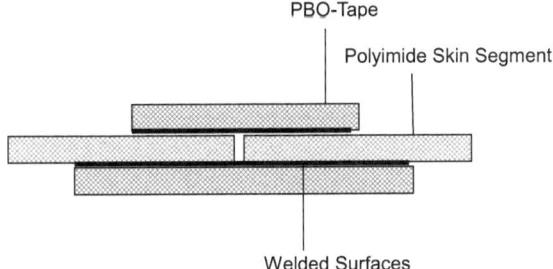

Fig. 7.66: Geometry of welded PBO-Polyimide two spliced lap-joint.

7.3.5 Ballute Instrumentation

It is desirable to equip the ballute with instrumentation, such as RF aerials and sensors. Several methods to bond such instruments to the ballute skin are shown in Fig. 7.67.

The simplest method is to place the sensor wire or antenna (idealized in Fig. 7.67 as a round electrical conductor) onto the ballute skin and fix it in place with an adhesive tape, perhaps the same as used in manufacturing the ballute seams.

Such was the instrument fixture on the MIRIAM ballute (see Fig. 7.68). The clear advantage is that the simple tape fixture requires comparatively little effort to manufacture. The disadvantage, however, is that large pockets of air get trapped between the conductor and the skin. These air pockets expand when ambient pressure drops, weakening the fixture and increasing stress. They also make seam fixtures more bulky than necessary.

An improved version of a tape fixture would wrap the tape around the conductor at regular intervals, as shown in the centre sketch of Fig. 7.67. That leads to a slight stand-off, but one that can fold sideways. Air pockets will still be unavoidable, but the entire design is less bulky, contributes less to the production factor and is more easily to fold up later. The disadvantage is a more difficult and tedious production method, which has to make sure that no adhesive surface parts are left uncovered, which would otherwise stick to overlapping skin areas during folding.

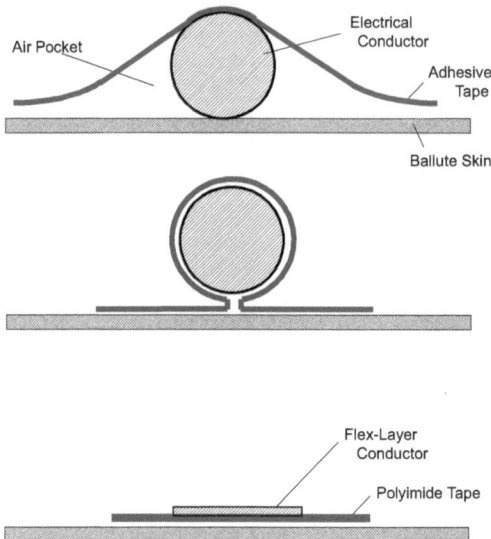

Fig. 7.67: Methods of bonding an electrical condictor to the surface of the ballute skin.

The ultimate in ballute-sensor incorporation would be an adhesive electronic flex-layer tape, with all the necessary conductors and shapes (i.e. antennas) etched into a conductive surface layer. The idea behind is that conductors are attached where needed, maybe even used as segment seams, but anyway simply glued in place like any other ordinary tape. In theory, such sensor types are possible, especially as substrates are normally polyimides akin to the UPILEX favoured for our ballute design. Such designs are not necessarily practical. For the time being, any flex layer design is limited in size to standard flex-layer board dimensions. While continuous production appears feasible, discussions with Lohmann Tapes of Neuwied, Metafol of Remscheid and Straschu Leiterplattentechnik of Oldenburg (all Germany) revealed that a development effort of around one to five million Euro is estimated to complete an assembly line for a mass-produced flex-layer adhesive tape.

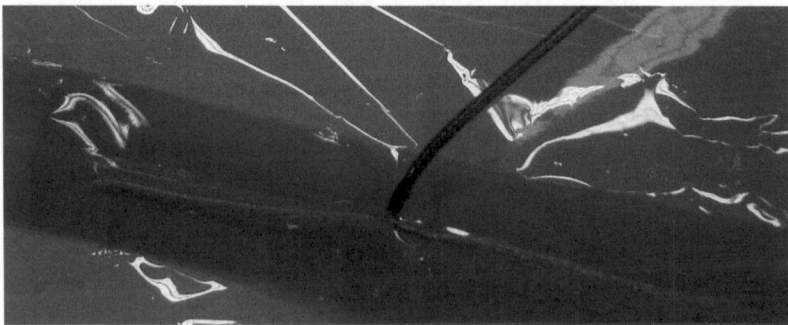

Fig. 7.68: The MIRIAM UHF antenna installed on top of a meridian seam.

The advantage of such a design would clearly be the absence of any inherent air pockets on a compact and easily foldable design, which is easy to incorporate into the ballute manufacturing process. If used as an anyway needed seam or reinforcement patch, the production factor can further be lowered, leading to an overall improved performance of the ballute design. Further analysis is necessary, if this design is the ultimate aim, investigating the behaviour of the conducting surface under strain, folding and under electrical load (EMC).

Whatever the design of the instrumentation, care must be taken when adding the fixture tapes. The thickness-change-in-steps rule from the shear lag theory (see chapter 7.3.4) should be kept in mind when attaching sensors. In addition to this, instrument fixture tapes should not hinder the ballute expansion, especially not locally, as this would add to skin pre-loading and the creation of a potential early failure point. On long wires, enough wire slag should be incorporated, so that the folding strain and thermal expansion strain in the ballute hull do not work against the stiffer wire. This would cause it to either brake or cause attachment fixtures to come loose and reattach uncontrolled.

Since any attachment protrudes from the surface, any sensor should be placed on the inside wall of the ballute, to keep the outside surface aerodynamically smooth.

The UHF antenna of the MIRIAM ballute spacecraft is shown in Fig. 7.68 and a global map of the ballute showing meridian seam and segment designations, along with thermistor and antenna locations, can be found in Fig. 7.69 [91]. Note that not all design rules were respected when designing the ballute instrumentation fixtures for MIRIAM, as some rules were results out of experience from building the MIRIAM ballute.

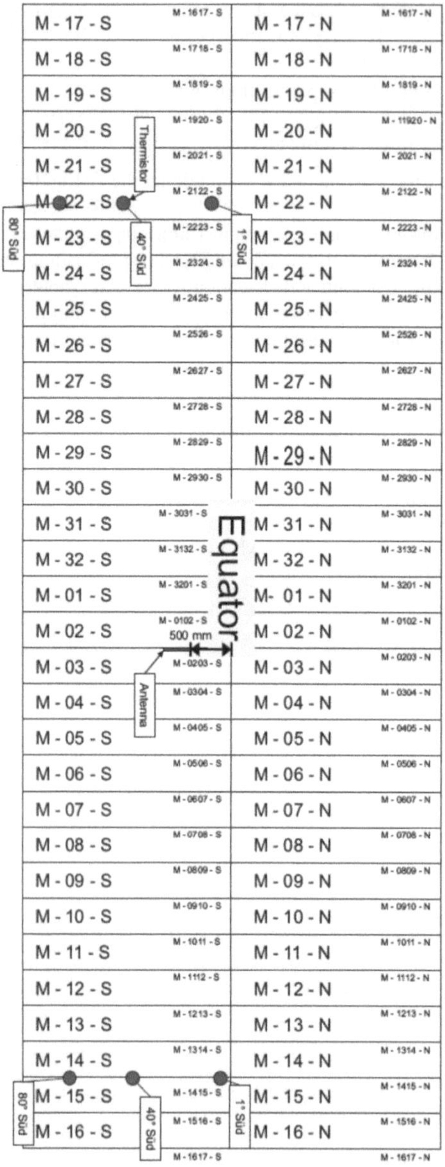
M - 17 - S
M - 1617 - S
M - 17 - N
M - 1617 - N
M - 18 - S
M - 1718 - S
M - 18 - N
M - 1718 - N
M - 19 - S
M - 1819 - S
M - 19 - N
M - 1819 - N
M - 20 - S
Thermistor
M - 1920 - S
M - 20 - N
M - 11920 - N
M - 21 - S
M - 2021 - S
M - 21 - N
M - 2021 - N
M - 22 - S
M - 2122 - S
M - 22 - N
M - 2122 - N
80° Süd
M - 23 - S
40° Süd
M - 2223 - S
1° Süd
M - 23 - N
M - 2223 - N
M - 24 - S
M - 2324 - S
M - 24 - N
M - 2324 - N
M - 25 - S
M - 2425 - S
M - 25 - N
M - 2425 - N
M - 26 - S
M - 2526 - S
M - 26 - N
M - 2526 - N
M - 27 - S
M - 2627 - S
M - 27 - N
M - 2627 - N
M - 28 - S
M - 2728 - S
M - 28 - N
M - 2728 - N
M - 29 - S
M - 2829 - S
M - 29 - N
M - 2829 - N
M - 30 - S
M - 2930 - S
M - 30 - N
M - 2930 - N
Equator
M - 31 - S
M - 3031 - S
M - 31 - N
M - 3031 - N
M - 32 - S
M - 3132 - S
M - 32 - N
M - 3132 - N
M - 01 - S
M - 3201 - S
M- 01 - N
M - 3201 - N
M - 02 - S
M - 0102 - S
500 mm
M - 02 - N
M - 0102 - N
M - 03 - S
M - 0203 - S
M - 03 - N
M - 0203 - N
Antenna
M - 04 - S
M - 0304 - S
M - 04 - N
M - 0304 - N
M - 05 - S
M - 0405 - S
M - 05 - N
M - 0405 - N
M - 06 - S
M - 0506 - S
M - 06 - N
M - 0506 - N
M - 07 - S
M - 0607 - S
M - 07 - N
M - 0607 - N
M - 08 - S
M - 0708 - S
M - 08 - N
M - 0708 - N
M - 09 - S
M - 0809 - S
M - 09 - N
M - 0809 - N
M - 10 - S
M - 0910 - S
M - 10 - N
M - 0910 - N
M - 11 - S
M - 1011 - S
M - 11 - N
M - 1011 - N
M - 12 - S
M - 1112 - S
M - 12 - N
M - 1112 - N
M - 13 - S
M - 1213 - S
M - 13 - N
M - 1213 - N
M - 14 - S
M - 1314 - S
M - 14 - N
M - 1314 - N
80° Süd
M - 15 - S
40° Süd
M - 1415 - S
1° Süd
M - 15 - N
M - 1415 - N
M - 16 - S
M - 1516 - S
M - 16 - N
M - 1516 - N
M - 1617 - S
M - 1617 - N

Fig. 7.69: Global map of the MIRIAM ballute [91].

7.4 Manufacturing Considerations

The manufacturing methods have to ensure that the ballute obtains the right shape, complies with planetary protection and space environment requirements (such as outgassing) and withstands all mission loads with sufficient margin. On top of all that, manufacturing methods have to be actually doable and practical. The manufacture of the ballute is therefore constrained by available materials, manufacturing tools and infrastructure.

All manufacturing methods presented herein have been jointly developed together with the Ballute-Team of the Mars Society Germany and were supported by the companies IABG at Ottobrunn, Lohmann Tapes at Neuwied, Technolplot at Ismaning and Die Firma at Nuremberg, all Germany [92].

7.4.1 Envelope Parts

When designing the ballute and its individual envelope parts, available cutting methods have to be kept in mind.

Any of the materials applicable to a high Mach number drag ballutes for entry applications are necessarily durable and tough. Naturally, they are just as tough and durable, when worked on with tools. Any variant of UPILEX turned out to be very abrasive on cutting tools, such as knives and scissors, causing excessive wear and quickly leading to a bad quality cut. Cuts of

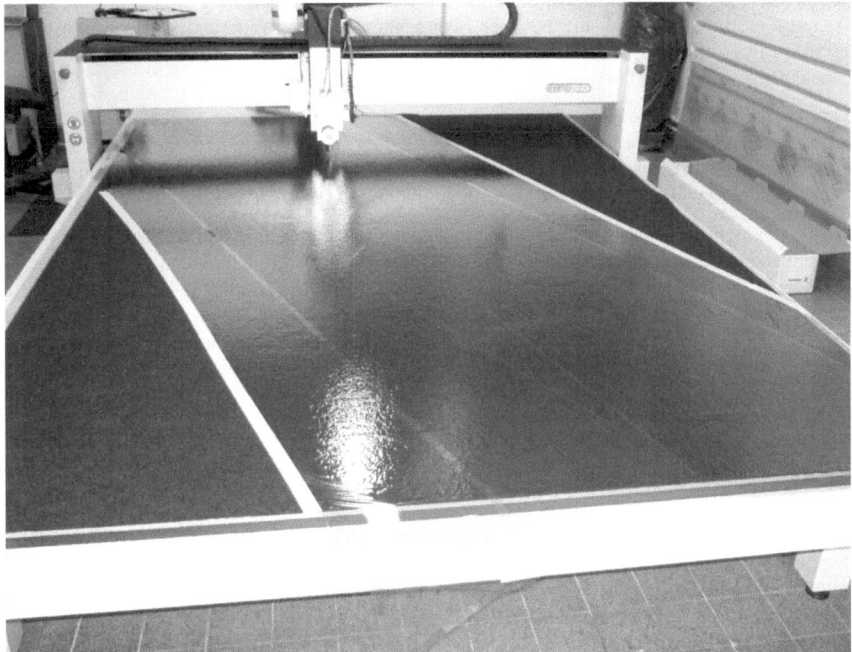

Fig. 7.70: Cutting out parts for the MIRIAM ballute on a cut-plotter with an oscillating blade.

bad quality exhibit little fissures and jagged edges, that will act as crack roots under tension. Segments must also be manufactured and assembled with great precision, since neither of the materials in question are very elastic, as compared to classic balloons. To exemplify the circumstance, one might consider the ultimate lower segment limit: a toy balloon. Made with no seams, the balloon obtains its shape through the interior pressure which it holds through the tension of its elastic rubber skin. Another good example would be a soap bubble. Neither have any significant production tolerances requirements, when it comes to segment shape. Polyethylene balloons and beach balls are in the middle somewhere. They usually have a smaller number of segments (8 maybe or even just 6) because it's close enough to a sphere for these elastic materials to stretch the rest. Because this stretching can also compensate for inaccurately cut and assembled seams, manufacturing accuracy requirements are low. Additional stress pre-loading of the skin, due to the added strain, is also hardly an issue.

With an UPILEX ballute in contrast, mission loads get close to material load limits, especially as a particularly light weight design is sought. Therefore, we must seek to minimize inherent load offsets through strain. Consequently, production tolerances shrink to the point where the manual cutting of the segments becomes impractical.

This challenge can be met with a flat-bed cut-plotter, that features an oscillating blade and a vacuum suction bed, to fix the film during cutting. Such devices are sold by the company Technolplot of Ismaning, Germany, and are available with beds of any length, which is a multiple of their 3 or 4 metre standard bed segment. Technolplot manufactured several test segments with their own demonstration cut-plotters. Tests with a rolling pin cut plotter, which is more commonly used in the advertising industry, were unsuccessful, because the comparatively thin ballute film cannot be transported by the plotter and always jams the mechanism, destroying the segment.

One of Techplot's customers, the advertising company Die Firma of Nuremberg, Germany, who has a suction bed cut plotter with the oscillating blade option and one three metre bed segment for a working table manufactured all parts for the MIRIAM ballute (see Fig. 7.70). The results were of very good quality, with no major problems encountered.

If this production option is chosen, a plot cutter has to be available that features a bed which is long and wide enough to accommodate at least one segment in length and width. In case of the beach-ball production pattern with an equatorial seam, the bed has to be large enough to comfortably accommodate a quarter circumference of the ballute. Since plot cutters can have any length, but are limited to their standard width, this particular method of designing and manufacturing the ballute limits the number of segments the ballute can be subdivided into (see also chapter7.3.2.4).

For ARCHIMEDES' 10m ballute, the bed would have to be at least 8 metres in length, plus the machine loss (the bed length which cannot be used, due to the minimum clearance that the cutting head needs to maintain to either rim). If the same unit as the one used for MIRIAM is chosen, at least three bed segments have to be used, to comfortably accommodate one segment. Since the standard bed width is 2 metres, the beach ball pattern of ARCHIMEDES has to have at least 18 segments around the equator. Since in chapter7.3.2.4 the minimum number of segments was determined to be 32, this is acceptable, actually allowing more than one segment to be cut out simultaneously.

7.4.2 Envelope Assembly

When assembling the envelope, we must bear in mind that the stiffer seams are rarely three dimensional objects. In case of a beach-ball pattern, seams bend around the centre of the spherical ballute, but they still only form a barrel segment, rather than a spherical surface. This means that additional strain must be expected in the skin around the perimeter of the seams. This problem becomes worse, when the seam which later forms a barrel is manufactured flat on a table, because the two skin segments which are joined at the seam already want to form a three-dimensional object.

Fig. 7.71: Assembling parts of the MIRIAM flight model ballute on a quarter circle support frame [92]. (MSD / IABG)

To minimize the additional strain during the manufacturing process, the MIRIAM ballute segments were assembled on a quarter circle support frame (see Fig. 7.71). The frame was manufactured from standard ITEM aluminium elements and has the same radius as the ballute. Therefore, the surface on which the seams can be manufactured already has the barrel shape of the seam and segments can adapt to their own barrel shape, by hanging loose.

Assembly of any space bound ballute requires a clean room qualified for space vehicle handling. All ballute materials in question are very good electric insulators, therefore accumulating a static potential that easily attracts dust, human skin and hair particles. These would not only violate planetary protection rules (the Mars-bound ballute has to be sterilized anyway), but fall on bonding surfaces, weakening the adhesive or heat bond and causing manufacturing imperfections.

For the same reasons, a clean surface is imperative. To enhance the bonding properties of the film, surface of the polymer has to be "activated", regardless of whether the bond will be adhesive or welded. Cleaning should be done with industrial alcohol that evaporates without leaving traces. The activation of the surface is achieved through corona treatment (see Fig. 7.72), whereby an electric field frees up loose polymer ends that can more easily interlock with the bonding layer.

The assembly of the MIRIAM ballute is described in great detail in [92].

As already discussed in chapter 7.3.4 adhesive tapes alone won't hold during the atmospheric entry of the ARCHIMEDES Mars ballute, so one of the welding methods have to be applied. For the assembly of the 10m ARCHIMEDES ballute, a welding machine of the type used by Ballonbau Wörner will have to be used. Since this machine runs along a linear track, the support frame used above cannot be used. If the machine cannot be transported to a large clean room facility, such as IABG's, that can accommodate the large ballute segments, the design and manufacturing of an appropriate purpose-built welding machine should be considered.

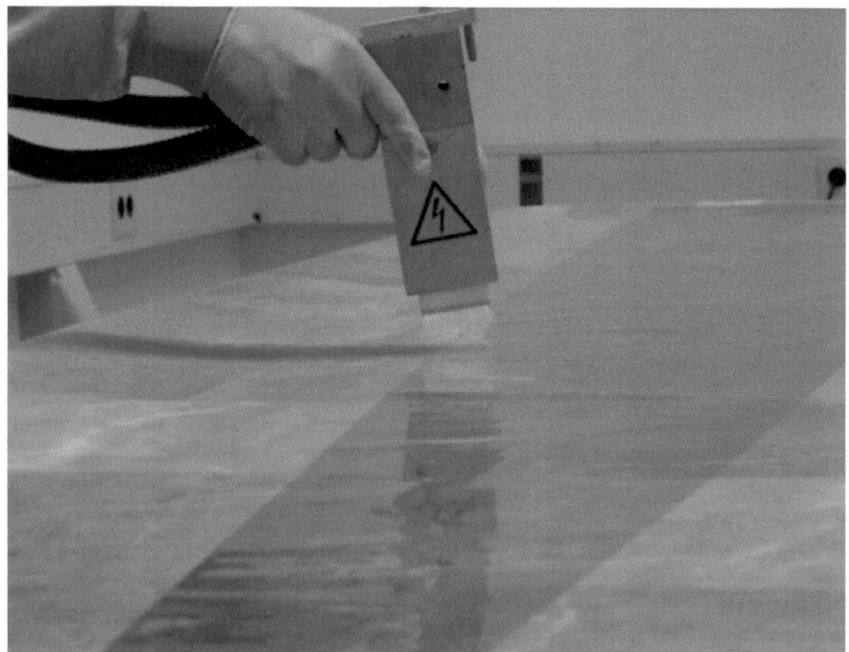

Fig. 7.72: Corona treatment of the skin segment to activate the polymer surface [92]. (MSD / IABG)

8 Transportation and Deployment System

8.1 Statement of Purpose

The transportation and deployment subsystem (TDS) has to hold and protect the ballute spacecraft from the point of integration to the ballute spacecraft's point of release. During interplanetary cruise, the ballute has to be kept well tempered and prevented from cold welding of packed envelope layers. Upon deployment, the system has to ensure that the ballute spacecraft properly unfolds in space and is held in position until fully inflated. Eventually, the system has to safely release the ballute spacecraft.

8.2 System Design

The TDS concept was developed in cooperation with the Institute of Light Weight Structures of the University of the Federal Armed Forces of Germany in Munich [94]. The system developed for MIRIAM is shown in Fig. 8.1. The TDS consists of a pyramidal container with a hexagonal base geometry (referred to as "the blossom") within a pyramidal and hexagonal structural frame. The blossom leafs are fixed to the blossom base with spring loaded hinges and rest against Teflon® bearings over which they can glide inside the structural frame. The

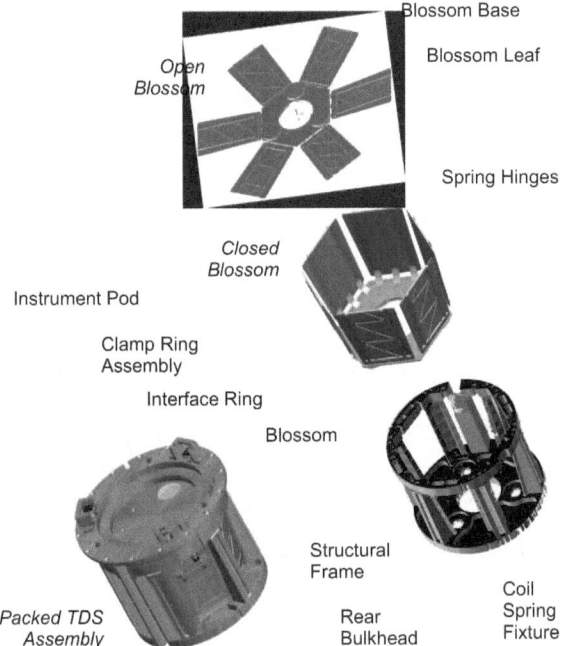

Fig. 8.1: Main elements of MIRIAM's transportation and deployment subsystem assembly [93].

blossom base is attached to the rear bulkhead with a set of conical coil springs and steel retention wires.

The ballute spacecraft is folded up in a hexagonal pattern with the instrument pod on top and the inflation hose interface section sticking out on the rear (see chapter8.4). The ballute rests on the blossom base and is fully enclosed by the blossom leafs. The instrument pod rim rests on the tip of the blossom leafs. A clamp ring assembly holds the packed ballute spacecraft and the blossom inside the main structural frame by holding down the instrument pod. The clamp ring is fixed to the structural frame with releasable bolts and a complement of separation springs.

8.3   Theory of Operation

8.3.1   General Principle

An artist's rendition of the deployment sequence of ARCHIMEDES (see also chapter10.3) is pictured in Fig. 8.2 [95]. The clamp ring assembly presses against the the ballute spacecraft, which rests against the blossom leafs and the spring loaded blossom base. If the clamp ring assembly fixtures are released, the separation coil springs push the clamp ring assembly away from the structural frame.

The conical coil springs underneath the blossom base push the blossom and ballute spacecraft outward, while the spring loaded hinges make the leafs fold away from the ballute spacecraft. The blossom is allowed to travel forward until the retention wires tighten. The ballute spacecraft will continue to be carried forward by its momentum, unfolding the inflation hose and part of the ballute package. The package and inflation hose will already unfold, through the rapid expansion of the air trapped inside the package (see chapter 8.4.2).

Once the ballute is fully inflated, the inflation hose clamp assembly will release the inflation hose, thereby freeing the ballute. The service spacecraft will then be removed from the ballute spacecraft by cold gas thrusters (see also chapter9).

The deployment mechanism was successfully tested on ESA's 40th Parabolic Flight Campaign and the REGINA space test (see chapters8.5.1 and8.5.2).

8.3.2   Protective Atmosphere

Trapping air inside the ballute during manufacturing and packing is unavoidable (see chapter8.4.2). At the same time a little gas can help preventing the folded ballute envelope layers from cold welding together.

On an interplanetary cruise lasting several months to years, cold-welding of materials under vacuum conditions is a serious threat. It is a bond that forms between two materials attached to each other in vacuum. It happens uncontrolled and does not have to be very strong, but can nevertheless hamper or even jeopardize successful deployment of the ballute. Trapped gas inside the ballute automatically helps to prevent that phenomenon to occur. To prevent the outside layers from cold welding, a slight atmosphere has to be added to the ballute container. A pressure of 1 hPa of Helium is considered sufficient and can be kept up by the IGSS auxiliary gas distribution system.

To soften the ballute skin for a smoother deployment, the TDS heating system may be designed such that it can preheat the ballute spacecraft to higher temperatures. In combination with a TDS flooding system fed from a tank elsewhere on the service spacecraft, this heating technique might also be used to bake a ballute dry and purging the moist atmosphere from the TDS blossom, in case a PBO-containing ballute is used (see chapter 6.2.2).

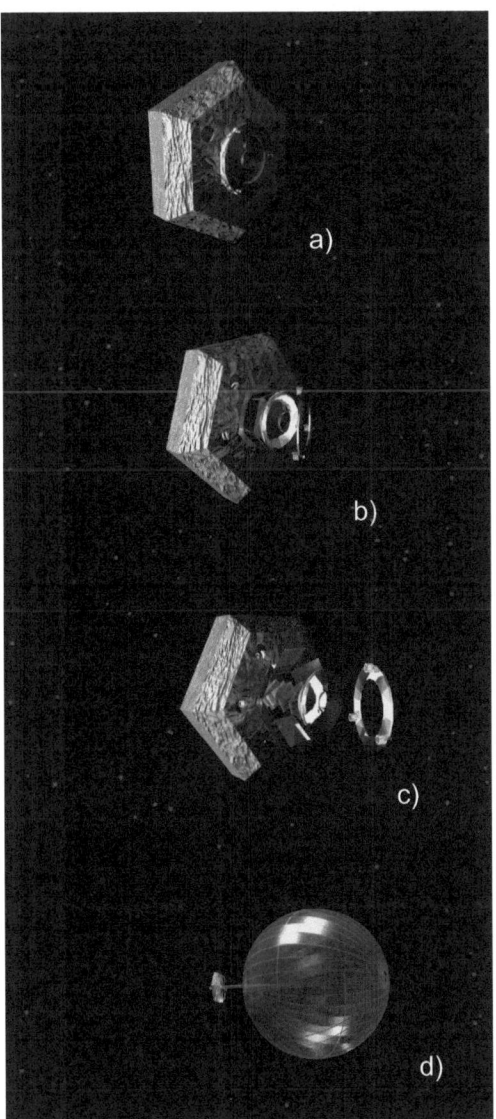

Fig. 8.2: Artist's rendition of the deployment sequence of ARCHIMEDES: a) Stowed configuration; b) clamp ring assembly separation; c) ballute package deployment; d) inflation (from [95]).

8.4   Ballute Packaging

8.4.1   General Principle

The ballute packaging technique was developed together with the Mars Society Germany (Mars Society Deutschland e.V,) and was coordinated with the development of the TDS. The principle underlying the folding pattern is shown in Fig. 9.3. All gores are folded over to one side along their centreline and the segment perimeter to form the shape of an E (when viewed head-on). The resulting package is then folded in a Z-fashion several times longitudinally and latitudinally. This results in a square base package.

The base package can then be brought into any desirable shape. In our case the base package is turned upright by 90° and folded into a a hexagonal shape. Because this is a rather complex folding pattern, that does not open so easily, a simple, Z-folded pre-deployment section underneath the package was added. It offers easy initial deployment for the trapped air to rapidly expand and ensures that the windsock is not trapped inside the more complex base package. This is necessary, because a trapped windsock might later hamper the inflation process. The south pole, with the instrument pod, is folded directly over the other end of the base package, so it can rest on top of the flower and keep the package under tension during cruise. Mechanical package pressure prevents vibration and other launch loads from shifting the package and tearing the skin near fixed joints and fittings.

8.4.2   Packaging Efficiency and Trapped Gas

For the design of the TDS it is necessary to know the size of the blossom, so we seek an estimate for the expected packaging volume. We find that mostly empirically, by comparing the quality of different packaging methods. To measure the quality of a packaging method, we define a packaging efficiency factor, which sets into relation the packaging volume and the expected material volume, as it is common practice for the packaging of inflatables [10], so that

$$\eta_{Pack} = \frac{V_{EnvMat}}{V_{Package}} \tag{172}$$

In our case, we obtain the approximate envelope material volume, by assuming that the entire mass of the ballute (without the instrument pod, but with all instruments incorporated into the envelope skin) has an average density similar to that of the main skin material. In other words, we compare the resulting packaging volume to the volume that a solid block of skin material of equivalent weight would have. Thus is the envelope material volume obtained through

$$V_{EnvMat} = m_H \cdot \rho_s \tag{173}$$

Packaging efficiencies of 30 to 40 percent are typical values for parachutes packed with a hydraulic press and deployed with a mortar. The packaging efficiency of the MIRIAM ballute reached 32 percent, by carefully folding the ballute with the help of a large frame and pressing the package gently.

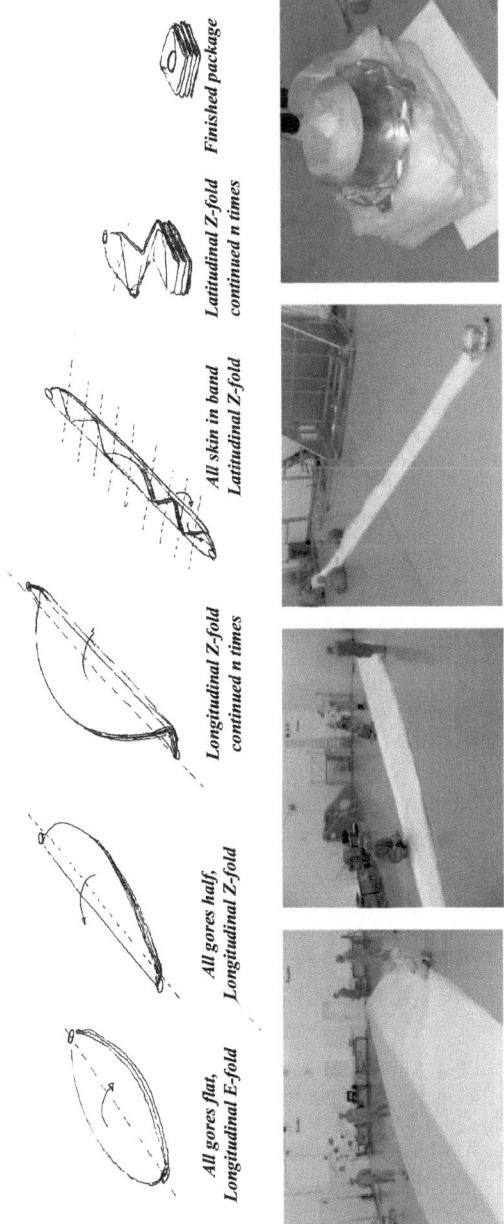

Fig. 8.3: Sketch of the ballute packaging principle and a test with a 13-m polyethylene test balloon.

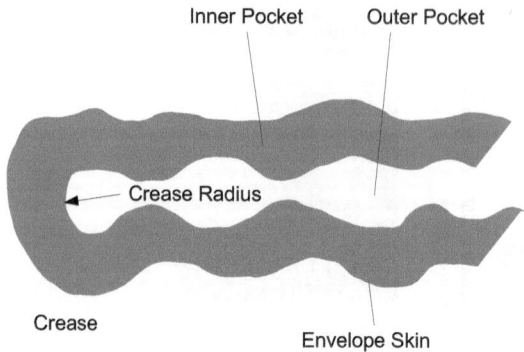

Fig. 8.4: Conceptual sketch of a close-up of a crease in the packed ballute envelope.

Naturally, the rest of the packaging volume is filled with air (Fig. 8.4). Those pockets of the volume which are outside the ballute envelope (outer pockets) will vent most of the air they contain under vacuum conditions. Those pockets which lie inside the ballute envelope (inner pockets) will retain the amount of air that was trapped. To get an idea of the amount of trapped air inside the ballute, we have to estimate a relation between the inner and outer air pockets.

From packaging experience, we know that a lot more air is trapped in outer air pockets than in inner air pockets. The reason is that the inner air pockets can be evacuated quite efficiently, by sucking the air out through the instrument pod fitting, while pressing down on the partly folded layers of the main E-folds and pushing resulting air bubbles towards the "exit" at the pod fitting. The bulk of air is trapped in outer pockets under subsequent Z-folds. Another reason for this fact is that the initial E-folds have a much smaller crease radius than subsequent folds, which have to deal with an already thicker package. The larger radius also encloses more space.

To sum up the above considerations, we assume that the air trapped inside the packed ballute envelope is less than half of the entire gas pocket volume:

$$V_{TA} \; < \; 0.5 \cdot \left(V_{Package} - V_{EnvMat} \right) \tag{174}$$

With equation (172) this becomes

$$V_{TA} \; < \; 0.5 V_{EnvMat} \cdot \left(\frac{1}{\eta_{Pack}} - 1 \right) \tag{175}$$

Let us briefly analyse the extremes of the above considerations: theoretically we reach the highest packaging efficiency by melting the ballute down and pouring it into a mould. The packing efficiency becomes 100% and ballute deployment becomes quite impossible, but no air would be trapped. We obtain the lowest packing efficiency by not packing the ballute at all and leaving it in its inflated, spherical shape. Note the packaging efficiency in this case is small, but not zero!

The amount of air trapped within the ballute envelope can now be calculated with the ideal gas law using atmospheric conditions at the time of packing. Because this is a coarse estimate, the trapped air can be simplified to have a density of 1 kg/m³, at the time of packing. For ARCHIMEDES' 10-m ballute, packed with an assumed efficiency of 30%, this amounts to less than 9g of trapped air. This is little more than 1% of the the roughly 800g of Helium with which the ballute would be inflated. If the ballute is left to expand freely under its own residual gas, the fully deployed ballute will have an interior pressure of less than 10 Pa.

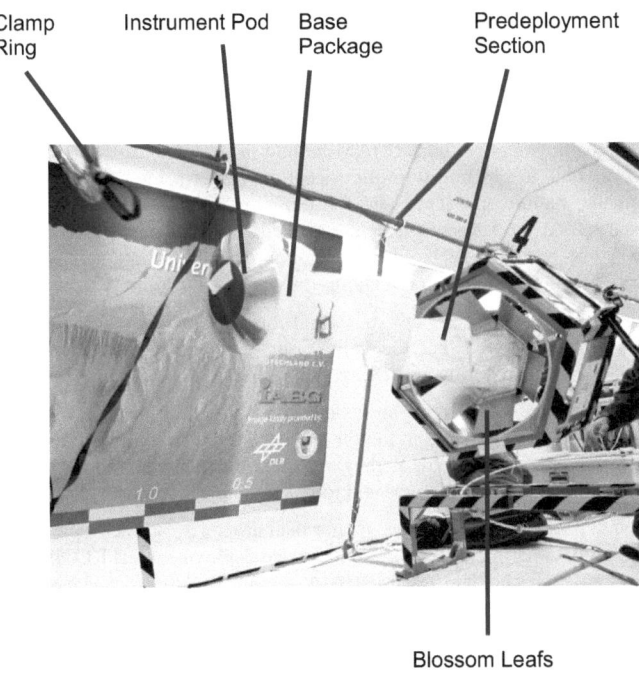

Fig. 8.5: *Unfolding ballute package in weightlessness, during parabolic flight testing (Photo: ESA).*

While trapped air is therefore negligible for the final ballute mission, it does have an effect on the deployment. Recall that it was packed at surface pressure and therefore has around 1000 hPa pressure when the package pops open. An unobstructed ballute deployment path is thus of the essence, so that the trapped air can expand quickly and freely, without exerting too much force on any single point on the ballute hull.

To ensure easy and rapid deployment of enough ballute volume to cause a significant pressure drop is the task of the pre-deployment section (see Fig. 8.5). At roughly one fourth to one third of the longitudinal E-package length, it offers enough easily deployable volume for the pressure to drop quickly below 100 Pa and be of no further concern.

The same set of equations can now be used when we seek an estimate of how much gas is needed during the entire transportation time. The maximum volume of the protective atmosphere can be found by assuming that no air at all is trapped in inner pockets:

$$V_{PA} < \left(V_{Package} - V_{EnvMat} \right) \tag{176}$$

Since a packaging volume pressure as little as 100 Pa is sufficient to prevent cold welding, the required amount is very small. If Helium is used on the ARCHIMEDES TDS, a total of 2g is sufficient. At pressures and densities this low, an initial charge of nitrogen or helium inside the ballute container can be used. Leakage at such low pressures is minimal, so the container doesn't even need a very sophisticated pressure vessel.

The deployment of a ballute in space, under the influence of internal trapped gas, can be seen in Fig. 8.6. It shows the deployment of the MIRIAM ballute in space. Because a stuck bolt had caused the service module to separate from the rocket only after some delay, the clamp ring was still retaining the ballute spacecraft and the blossom, when the inflation control command sequence started. At the subsequent deployment of the ballute, an estimated ten percent of the inflation gas was already building up pressure in the inflation hose clamp assembly, causing a rapid deployment, when the clamp ring eventually came loose and freed the ballute.

This unintentional experiment showed that the ballute can deploy under a much higher interior pressure than the air trapped during packaging would create.

In Fig. 8.6, the instrument pod is out of frame and above the image's top frame. The service module and the north pole of the ballute are in the far left, also out of frame. Note the hump of the base package and the remaining E-folds as they open up. The sequence was captured by one of the service module's on board TV cameras and took less than a second.

It is important to note that packaging efficiencies greater than around 25%, such as it was the case with the parabolic flight experiment (see 8.5.1), the space deployment test REGINA (see chapter 8.5.2) and finally the space flight test MIRIAM, require that the package is placed inside the TDS blossom and allowed to sit for at least 12 hours, before the rest is packed and locked in place for the flight. The wait time allows the package to adapt better to the shape of the blossom, reducing strain inside the layers.

Base
Package

E-Folds

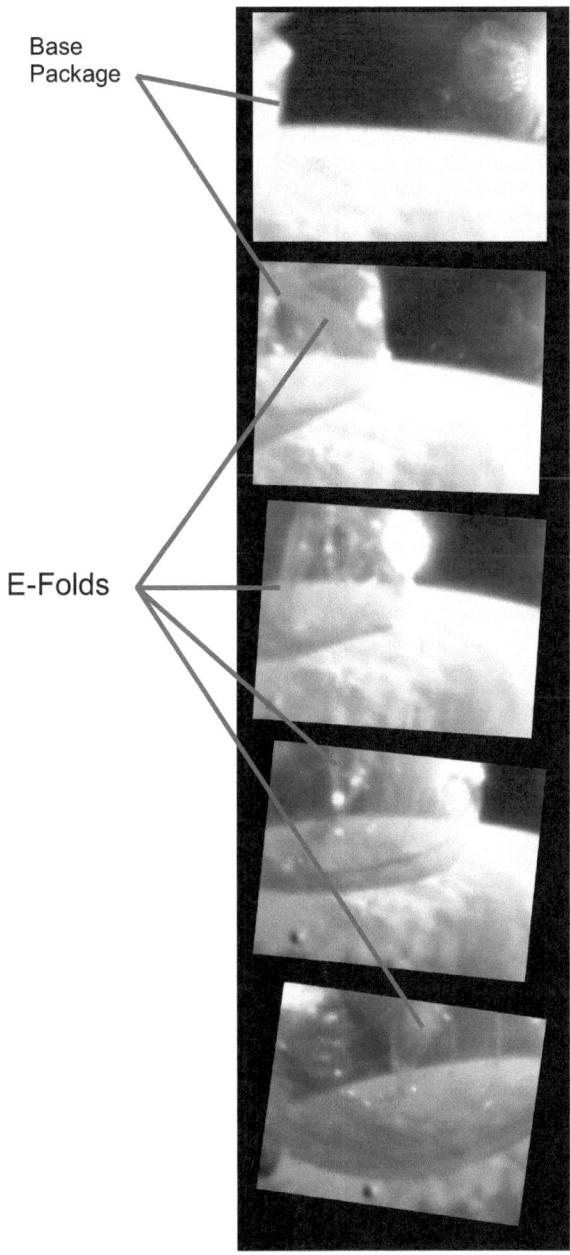

Fig. 8.6: Deployment sequence of the MIRIAM ballute in space.

8.5   Deployment Tests

8.5.1   Parabolic Flight Deployment Test

To test the TDS for functionality and obtaining data with which to correlate mathematical predictions of the balloon deployment speed, weightlessness is necessary. Weightlessness is available on parabolic flights. This also allows simulating various failure modes and testing the packaging method for its unfolding characteristics (see Fig. 8.7). Originally, two campaigns were planned to test a scale model of the deployment system and later an improved 1:1 version, dubbed the "ARCHIMEDES Release Mechanism" (ARM) test. The first of these campaigns, ARM Stage 1, was successfully flown in June 2005 on ESA's 40th parabolic flight campaign in Bordeaux. The second one had to be cancelled, due to a scheduling conflict with the REXUS 3 sounding rocket flight and as of now remains to be done. However, the results demonstrated the suitability of the deployment method very nicely [96]. Experience and insight gained during this campaign also allowed continuing the development of hardware for space tests.

The actual test article was a 1:2 scale model of the mechanism and the ballute for ARCHIMEDES [97]. It consisted of the hexagonal central structural box with the blossom container and the test balloon, the release mechanism, the clamp ring that locks the blossom container in place, a mass dummy for the space probe's instrument pod and a hexagonal safety frame that enclosed the entire working volume of the test mechanism for easier handling and greater safety (see Fig. 8.7).

The hexagonal safety frame structure was 600mm from wall to wall and 700mm diameter including handles. It was around 340mm long and the entire structure weighed approximately 15kg.

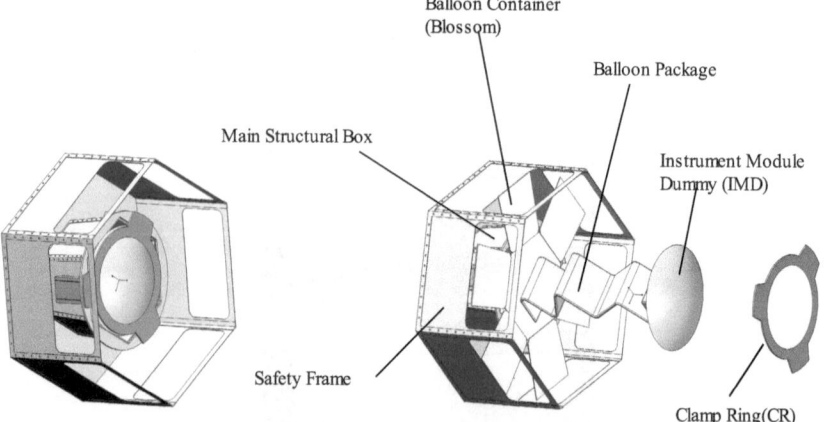

Balloon Container
(Blossom)

Balloon Package

Main Structural Box

Instrument Module
Dummy (IMD)

Safety Frame

Clamp Ring(CR)

Fig. 8.7: Conceptual sketch of the flight test article for parabolic flight testing

Depending on the test mode, the test system was either floated freely inside the experiment volume or fixed to its stowage table [98]. The model space probe was pre-deployed during the μ-G phases of the parabolic flight, whereby the pre-deployment package opened fully, but the main package remained secured by PE-film in all but the very last parabola, after which the more complex main package could not be repacked without the elaborate folding rig to obtain the necessary packaging efficiency. During level flight between parabolas, the system was normally repacked within less than 2 minutes. The balloon was not inflated.

The test setup included:

1. The actual test system, scale 1:2, with a safety frame;
2. 4 Video cameras and mounting brackets;
3. 2 remote control systems (one flight and one flight spare unit);
4. A background image for each flight day, to provide visual reference for the deployment process analysis.

The system deployed the balloon repeatedly, under various conditions and failure modes, without malfunction. It was observed how the balloon deployed and behaved in all cases. The separation speed of the deploying balloon was measured using video analysis and a scale. This information is important, to design the separation springs and for the timing of the IGSS command sequence and other events in the deployment sequence.

Video analysis of the tests showed that the predicted deployment speed and balloon behaviour were close to expectations. However, minor adjustments were made to improve the deployment behaviour.

Results of the campaign were used in the design of the two space flight tests REGINA and MIRIAM.

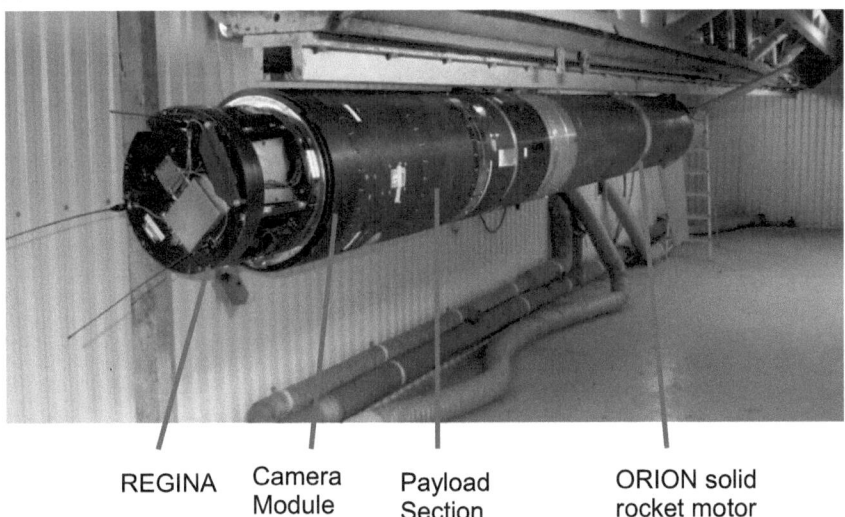

REGINA   Camera   Payload   ORION solid
          Module   Section   rocket motor

Fig. 8.8: The REXUS-3 rocket hanging on the launch rail with the nose cone removed, showing REGINA mated to the Camera Module.

8.5.2   REGINA In Space Deployment Test

The available cabin dimensions on board commercial jet liners, as well as the presence of an atmosphere, do not allow deployment tests of the full ballute, under the influence of residual trapped air on parabolic flights. The relatively short duration of weightlessness (only 20 seconds) also requires additional tests. Therefore, a test must be made in space, to observe the behaviour of the TDS design and the deployment of a dummy ballute, without the restrictions of a parabolic flight. It can, however, be made only once and is more difficult to observe. Such a test was made under the name of REGINA and was completed in just 8 months. It was flown on the REXUS-3 sounding rocket on April 5, 2006 (Fig. 8.8). To observe the deployment, a dedicated Camera Module inside the rocket interface ring was used as an observation platform, while marker lights on the flight system were to allow the tracking of movements, through photogrammetric analysis [99]. The test was done in collaboration with the Mars Society Germany, the Institute of Light Weight Structures and the Institute of Photogrammetry and Cartography at the University of the Federal Armed Forces of Germany, in Munich. A single stage Orion solid-fuel rocket took the REGINA flight system to the edge of space at 93 km, on a ballistic trajectory. The rocket was sponsored and operated by the MoRaBa group of the DLR space operations centre in Oberpfaffenhofen and the Swedish Space Corporation.

Fig. 8.9: The REGINA flight system as filmed by the Camera Module.

Technically REGINA was essentially a repetition of the parabolic flight test, with an autonomous system under near-space conditions, testing the behaviour of the ballute package under the influence of trapped gas. REGINA comprised the TDS with a flight computer that operated the release mechanism. The instrument pod carried no instruments and only acted as a mass dummy. The REGINA TDS structural frame too had marker LEDs around its perimeter. The intention was to track the LEDs as marker points on videos recorded by 10 solid state video camcorders (VCC) with VGA resolution. Images show that, despite the application of ultra-high-power LEDs, the contrast in these pictures is so strong that the Marker points were hard to track on the dark structural frame (see Fig. 8.9 and Fig. 8.10a) and impossible to make out on the shining aluminium of the instrument pod (see Fig. 8.10b). Later lab tests with the finished Camera Module system [99] also showed that videos of a much higher resolution are recommended for tracking, as the VGA-resolution (640x480 pixels, the highest which was then commercially available) of the installed VCCs was acceptable for that task, but not very well suited. Marker light tracking was therefore not installed on the later MIRIAM test. For future tests, other means of motion analysis have to be used.

Due to an erratic timer setting, the rocket's payload section collided with the REGINA flight system shortly after the latter was separated. Images such as Fig. 8.9 obtained by the Camera

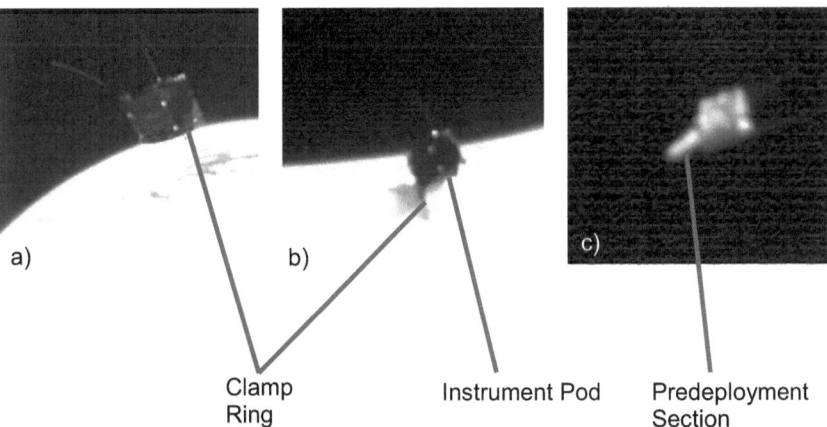

a)   b)   c)

Clamp   Instrument Pod   Predeployment
Ring   Section

Fig. 8.10: Deployment sequence of REGINA in space, as it drifts away from the rocket's payload section: a) before deployment; b) clamp ring rotating away from the system (blurred); c) pre-deployment section extending away from the blossom leafs (Images: Fleischmann / TU München).

Module and the T-REX fish-eye camera experiment, built by the Technical University of Munich, showed that the system deployed the ballute dummy correctly during the flight (Fig. 8.10). No significant inflation under residual gas could be observed. The results were deemed good enough to decide against a repetition of the test, in favour of a continuation of the flight testing programme.

The REGINA system was found by locals and returned to the Institute of Space Technology of the UBW for analyses. Airflow heating marks showed that the attitude of the flight system during its drop through the atmosphere was not straight with the blossom facing up (as it would be expected because blossom leafs act as flaps that increase drag while the battery-laden bulkhead pulls the centre of mass to the opposite side), but slightly skewed, hinting at a skew blossom deployment. Marks on the blossom leafs, as well as the still intact mechanism also showed that the blossom was in fact slightly skew and did not travel forward to the full extend of all retention wires, which is also consistent with the slightly skew attitude of the ballute dummy, with respect to the structural frame seen in Fig. 8.10c. The reason was found to be a retention wire that could not extend fully, because it was jammed at one of the coil spring base plates (it was still jammed when the system arrived back at the institute).

While this did not hinder correct deployment of the system and would most likely not have hindered a correct deployment of a real ballute, the retention wire system was redesigned to prevent such jamming. A separate bay milled into the aluminium bulkhead was incorporated into MIRIAM's design to stow the retention wires without the risk of jamming.

9 Inflation Control and Gas Storage System (IGSS)

The inflation control and gas storage system (IGSS)must handle the inflation gas and the inflation process from initial tank pressurization through interplanetary cruise up to the full and complete inflation of the ballute.

In case of a Mars Mission (see chapter 10.3 starting on page 208) the ballute can be a delicate, gossamer structure, before it is stiffened by internal pressure. We therefore seek precise control over the inflation hose pressure. Naturally, the system should be lightweight and small. For reliability reasons, it should have as few moving parts as possible.

9.1 System Design

The main problem of inflating a light, gossamer structure is the fact that high pressure gas from storage tanks or a gas generator has to be regulated and transferred to a low pressure inflation envelope, in order to prevent inflation hull rupture. In case a controlled system is chosen, a set of pressure reducers, regulators, throttles and valves must be used.

To decide which gas storage method is favourable, the three methods outlined in chapter 2.2 are qualitatively compared to each other in Table 10.1 and evaluated for use on our Mars mission.

	Compressed Gas	Gas Generator	Cryogenic
stores Helium or Hydrogen	Yes	No	Yes
storage time	Several years demonstrated for He	Several decades demonstrated	Several weeks demonstrated
energy consumption	During inflation	Triggering only	Permanent
storage loss	minimal with He and H_2	Almost none	high through boil-off requirement
thermal constraints	minimal	minimal	stringent
flight heritage	Yes	Yes	Yes
controllability of the inflation process	very precise	very imprecise	very precise
Predictability of gas composition	very good	very low	very good
Quality and purity of inflation gas	very high	very low	very high
system complexity	medium	low	very high
relative volume	high	low	medium
relative system weight	medium	low	high
Gas available for other systems during flight	Yes	No	Yes

Table 9.1: Decision matrix for inflation gas storage options

Since special emphasis has been put on the application of an atmospheric sounding balloon for Mars, a gossamer film ballute, filled with Helium, was found to be the most desirable configuration. Therefore, Helium has to be stored. Additionally, good controllability of the inflation process is desirable, to maximize the chances of successful deployment. Lastly, the balloon container should be kept slightly pressurised, to minimize the risk of cold-welding the balloon skin.

Based on these considerations, the implementation of a compressed gas system was chosen. A contributing factor to this decision was the consideration that an undoubtedly more complex mother spacecraft may require pressurized fuel tanks. In such a case, system trade-offs reveal that in most cases, a common gas handling system is a favourable option, somewhat compensating for the added mass and largely compensating the increased complexity, as compared to a gas generator system. So instead of having a compressed gas system and a gas generator cartridge, the gas handling system gets a larger tank and another branch leading to the inflation system.

Choosing a pressurized gas inflation system necessitates a pressure reduction stage and the desire to control the inflation pressure requires a system to influence the gas throughput. Standard pressure reducers are inappropriate for space applications because of their high part count, bulkiness and, in general, because of their various possible failure modes. Space qualified pressure reducers may exhibit a higher degree of reliability, but are still complex, bulky and on top of everything else, rather costly.

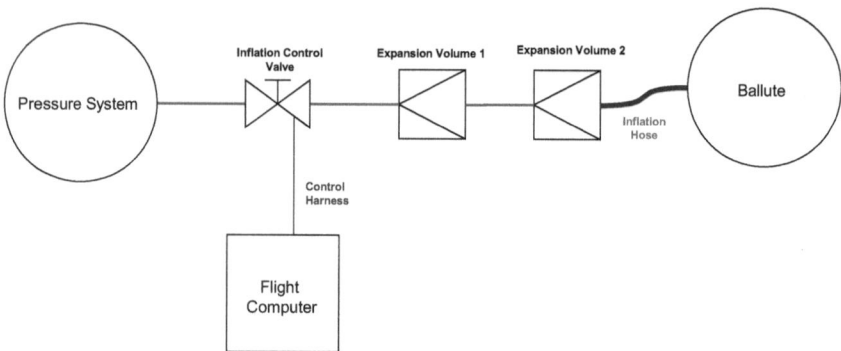

Fig. 9.1: Principle inflation control system concept overview showing all major components.

The inflation control concept developed for the task at hand only uses one single solenoid valve, as shown in Fig. 9.1, called the inflation control valve (ICV). It is the inflation control system's core functional unit and connected to two successive chambers of expansion volumes (the expansion chamber assembly). Upstream the ICV is connected to a pressure feed system. The two expansion chambers are separated by a pin-hole throttle. Following the expansion chamber assembly is the inflation hose and ultimately, the ballute.

Based upon this concept, a standard reference design for the IGSS was developed, which is shown in Fig. 9.2 and Fig. 9.4. The reference design of the IGSS has two parts. One part is on the side of the mother spacecraft and has mostly rigid, metal components which are separated from the ballute after the inflation process ends. This part is shown schematically in Fig. 9.2.

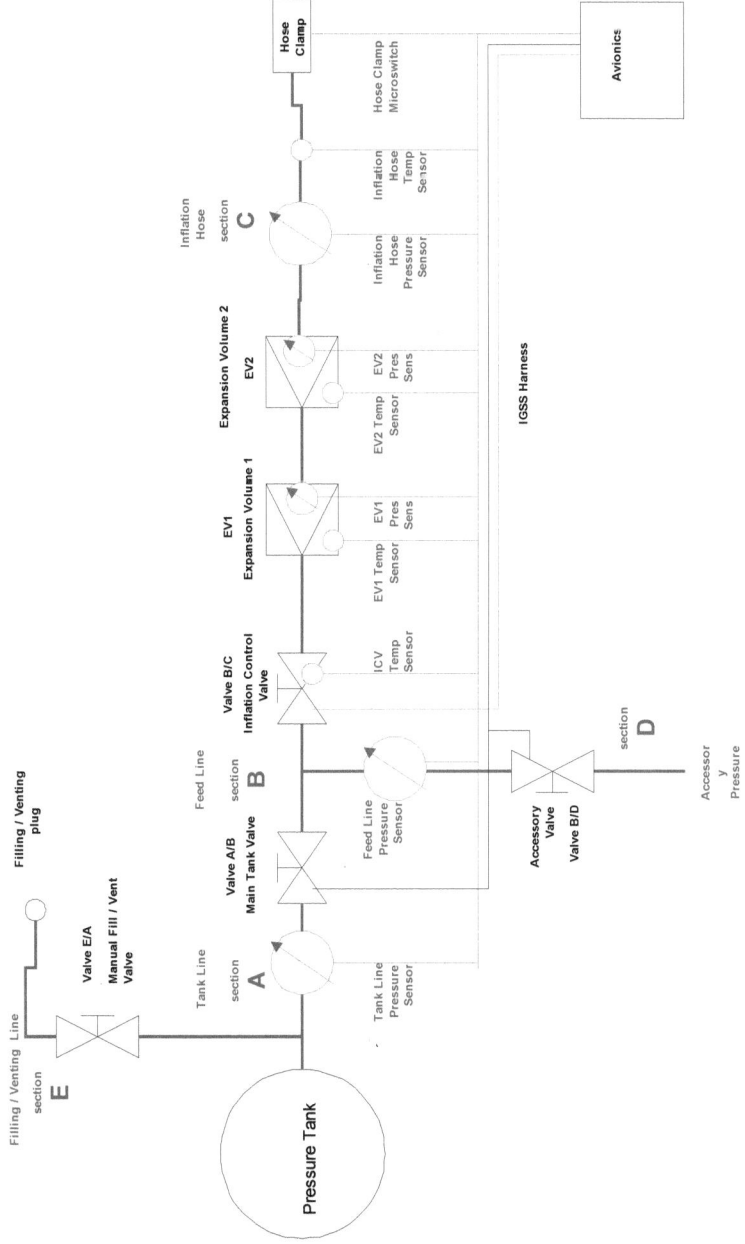

Fig. 9.2: Principal IGSS schematic, showing basic elements and suggested sensors.

The other part is built into the ballute itself and consists of soft, thin film parts. This second part is not separated from the ballute, but an integral part of it and stays with it until the end of the mission. This part is shown in Fig. 9.4.The inflation gas is stored at high pressure in the tank system. The tank system is sealed by the main tank valve and the tank pressurization and vent valve. This valve is used to pressurize or vent the tanks on the ground and is sealed after the final preflight pressurization. The main tank valve connects the tank system to the feed system, which directs high pressure gas to the inflation control system and other possible auxiliary systems. Auxiliary systems, such as the bleed system to maintain a protective atmosphere in the ballute container of ARCHIMEDES or the separation thrusters on MIRIAM, can be very mission specific and are not further discussed in this chapter. The inflation control system connects to the hose clamp assembly.

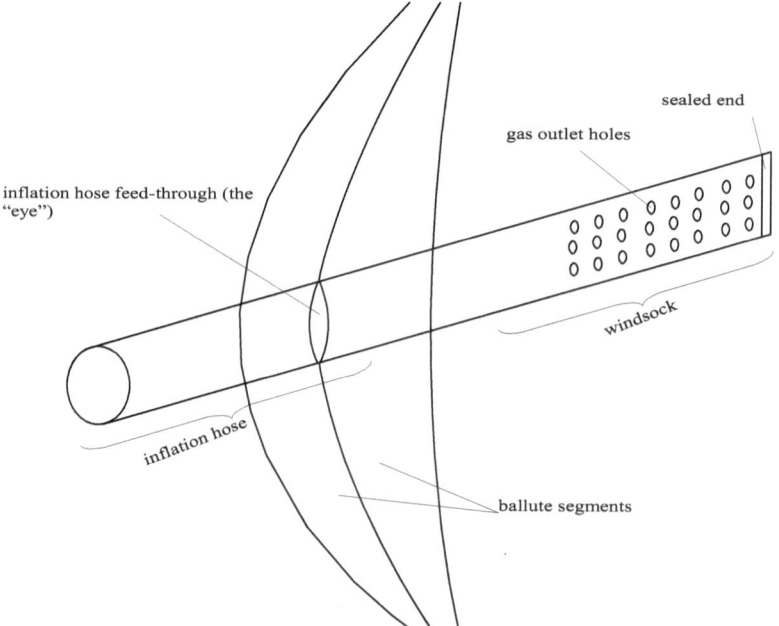

Fig. 9.3: Inflation hose and windsock on the ballute.

The inflation hose clamp assembly is the interface to the ballute part of the IGSS, as it connects to the ballute's inflation hose. This hose passes through the ballute's hull at a feed-through hole referred to as "the eye". At a distance away from the hull inside the ballute, the hose connects to the windsock.

The main advantage of this design is its mechanical simplicity. There is no need for a mechanically complex and relatively unreliable pressure regulator. Not counting the main tank valve (which is not strictly a part of the inflation system, but seals the tanks from the entire gas distribution system), all that is required is one solenoid valve. In theory, a system is even possible that has no need for a main tank valve, using the ICV to seal the tank. This, however, prohibits the use of the gas elsewhere on board and is a single point failure mode for inadvertent ballute package pressurization.

9.2   Theory of Operation

9.2.1   General Overview

The inflation control and gas storage subsystem is a compressed Helium system with dynamic pressure control. High pressure inflation gas from the tank system is directed through the main tank valve to the feed system, which distributes the gas to other systems and the inflation control valve (ICV). The ICV is a high-pressure solenoid valve which can open and close fast enough to inject an amount of gas into the first expansion chamber, causeing a pressure drop between 1:10 and 1:20. By controlling the opening time and frequency of the ICV, the flight computer can control the amount of inflation gas injected into the first chamber of the expansion chamber assembly, effectively controlling the first pressure reduction factor through pulse width and frequency modulation. From there the gas flows through a pinhole throttle into the second, larger expansion chamber. The pinhole throttle functions as a second pressure reduction stage and prevents pressure waves from the ICV to propagate downstream. Pressure wave propagation is further hampered by a set of deflector plates within the second expansion chamber.

Low pressure from the second expansion chamber flows through the inflation hose clamp assembly into the inflation hose and further into the windsock. The windsock finally distributes the inflation gas pressure equally inside the ballute through perforated walls, while the end is sealed. This prevents the inflation hose from vibrating and maintains a constant overpressure inside the hose. This constant overpressure keeps the flexible hose mechanically stable and the feed-through hole in the ballute skin open. The flow from the windsock into the ballute is the final pressure reduction stage.

Through this mechanism, the flight computer can effectively control the inflation hose pressure and gas throughput into the ballute, by a pulse-width modulation of the ICV. The commands for the inflation sequence are stored in the Inflation Control Command Sequence (ICCS) list in the computer's memory. In a simple system, ICCS entries are command sets containing pulse width, frequency and the number of pulses. These command sets have to be determined before the inflation process starts. ICCS commands my be determined by assumptions, simulations and experimental measurements in the lab. Such a system is simple, reliable and contains no more mechanical parts than a fully passive system with one ordinary solenoid valve. However, it is also rather imprecise and cannot adapt to variations in tank pressure, gas temperature or unforeseen situations such as delayed deployment or a false estimate of residual gas within the ballute package. In a more sophisticated system with a closed control loop, ICCS entries would be command sets with inflation hose pressure and time-line information, whereby the computer calculates actual gas pressure and throughput in real time, based on real time measurements. Such an elaborate system offers precise control over the ballute's gas content and hose pressure, but requires a complex control law and reliable sensor information. It therefore comes with all the design and reliability problems of closed real time control loops.

If left uncontrolled by simply opening the ICV, the expansion volumes would already cause a significant reduction in pressure through the pinhole throttles. However, a high pressure wave would propagate downstream and into the ballute, which would have to withstand a shock wave. Then, as the tanks blow down, the system would inflate ever more slowly.

To separate the ballute from the mother spacecraft, the inflation hose clamp opens to free the inflation hose, which can then slide from the assembly. Immediately after the clamp opens, pressure inside the inflation hose drops to zero. The hose and windsock and with them the feed through hole in the ballute skin collapse, thus effectively sealing the ballute.

It should be noted that the windsock method and feed-through hole are state of the art balloon technologies and were simply adapted to the application at hand.

9.2.2   Mathematical Description and Simulation

After consultation with the Institute of Fluid Dynamics at the University of the Federal Armed Forces of Germany in Munich [79], it was decided to use a mathematically most simple ideal gas model, as a first step in simulating the gas and its flow through the system. Its results were to give ballpark figures to assist the dimensioning of a test model. If necessary, a CFD simulation of the gas flow would be added at a later stage. A Matlab/Simulink model (see Fig. 10.16) was thus created and used for the determination of principle design parameters [100]. When the test system was later built, experimental results from it matched the simulations so remarkably well that a more elaborate model taking real gas effects into account was deemed unnecessary [101].

The simple, ideal model does not account for wall friction, turbulence or buffeting. However, it does take backflow into account, since the model calculates flow from high to low pressure areas. So any pressure building up downstream has an effect upstream. The equation system in Simulink is solved numerically with Simulink's built-in Runge-Kutta solver.

So to model the state of the inflation gas within any component of the system, the equation of state for ideal gases was used:

$$p \cdot V = m_g \cdot R \cdot T_g \qquad (177)$$

The prevailing gas density is defined by

$$\rho = \frac{m_g}{V} \qquad (178)$$

R is the gas constant specific to the inflation gas mixture and may be determined from the universal gas constant and the molecular weight of the inflation gas mixture. In case of pure Helium, this is:

$$R_{He} = \frac{R_{uni}}{M_{He}} = \frac{8.3145}{4.0026 \cdot 10^{-3}} \left[\frac{J}{kg \cdot K} \right] = 2077.27 \left[\frac{J}{kg \cdot K} \right] \qquad (179)$$

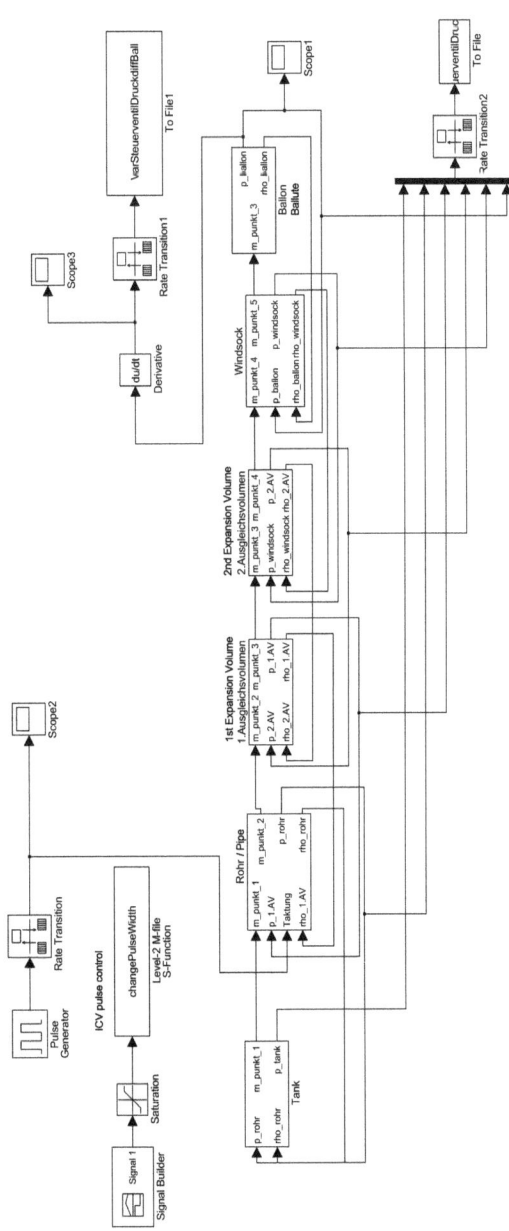

Fig. 9.4: The top-level of the simple IGSS simulink model [101].

Since the model assumes a strictly reversible (isentropic) and adiabatic behaviour, meaning that the exchange of energy across its boundaries is not permitted, the specific heat capacities c_v and c_p are independent from the gas temperature T_g and the ratio

$$\gamma = \frac{c_p}{c_v} \tag{180}$$

between the two is constant. Therefore, with

$$p \cdot V^\gamma = const \tag{181}$$

and (177), isentropic relations govern the pressure, temperature and density relations between two points:

$$\frac{p_2}{p_1} = \left(\frac{T_2}{T_1}\right)^{\frac{\gamma}{\gamma-1}} = \left(\frac{\rho_2}{\rho_1}\right)^\gamma$$

$$\frac{T_2}{T_1} = \left(\frac{p_2}{p_1}\right)^{\frac{\gamma-1}{\gamma}} = \left(\frac{\rho_2}{\rho_1}\right)^{\gamma-1} \tag{182}$$

$$\frac{\rho_2}{\rho_1} = \left(\frac{T_2}{T_1}\right)^{\frac{1}{\gamma-1}} = \left(\frac{p_2}{p_1}\right)^{\frac{1}{\gamma}}$$

Density and pressure within a flowing gas are given by the Bernoulli equation, which states that the sum of the static pressure p and the dynamic pressure q in a flowing medium are constant:

$$p + q = p_{total} = const \tag{183}$$

The dynamic pressure q is given by

$$q = \frac{\rho}{2} u^2 \tag{184}$$

where u is the velocity of the gas. To determine the initial velocity of the gas flow at the tank system connector, point 1 is defined to be inside the tank system and 2 is defined to be in the feed system behind the main tank valve. The volume of the solenoid valve chamber is neglected and the flow velocity within the tank assumed to be $u_1 \ll u_2$ so that

$$p_1 = p_2 + \frac{\rho}{2} u_2^2 \tag{185}$$

and

$$u_2 = u_{outflow} = \sqrt{\frac{2(p_1 - p_2)}{\rho}} \tag{186}$$

Of course, the assumption that differences between the gas density in the feed system and the tank system are negligible only holds if the solenoid valve has been open long enough to allow the gas pressure and densities to equalize. So the initial pressure wave propagation, just shy of the moment where the solenoid valve lifter hits the sleeve, cannot be modelled this way.

The mass throughput can be obtained from the continuity equation, which generally states that the rate at which mass enters a given control volume must be equal to the rate at which it leaves this volume:

$$\frac{\partial \rho}{\partial t} + \nabla(\rho u) = 0 \tag{187}$$

In the case of the inflation gas flowing through the inflation system pipes, the mass flow can be obtained from (187), by adding the cross section of the pipe and assuming a stationary flow for the simulated time step:

$$\dot{m}_g = \rho \cdot u \cdot A = const. \tag{188}$$

where $\dot{m}_g$ is the mass flow of the gas and A the cross sectional area of the flow volume.

The above mentioned set of equations are evaluated for each time step for the tank system, feed system, expansion chamber assembly, throttles, inflation hose and windsock and the ballute. As for the ballute, the flow velocity within it is again assumed to be much smaller than in the windsock, so that $u_{ballute} \ll u_{windsock}$. Solenoid valve chamber volumes and cross sections, interface clamps and other connections are neglected, because they are small as compared to the other control volumes.

The top level of the Simulink model of the IGSS can be seen in Fig. 10.16. A more detailed description of the model and its various subsystems and functions may be found in [101].

10 Mission and Spacecraft Design for Ballute Applications

The peculiarity of ballute missions is that, more than on any other space mission, the design geometry and mass of the space vehicle are interdependent on mission design and vice versa. This is true even when compared to ordinary entry vehicles, because the deceleration altitude itself is of no concern, so long as no one gets hurt, and because relatively dense and small vehicles can tolerate a larger variation in the ballistic coefficient with comparatively little impact on the trajectory.

Ballutes on the other hand, by their very purpose, ought to be large and light. Variations in mass have a more drastic effect on the ballistic coefficient and thus the trajectory design. If the trajectory has to meet more stringent requirements than the simple survival of the vehicle and its payload, the mission and spacecraft design process become more integrated activities.

This chapter will present a principal mission and spacecraft design method as well as designs made following this method. This includes the ARCHIMEDES Mars mission, which is the main purpose of the entire study, but also a test vehicle (MIRIAM) and other possible ballute applications.

10.1 Principal Mission and Spacecraft Design Guide

Presented herein is an easily applicable reference guide for practical mission design with ballutes based on the previously explained theory. The method described herein is intended to obtain preliminary design factors and order-of-magnitude figures that can later serve as a baseline for a more detailed design study. This method was developed and applied to the design of the missions presented in this chapter.

For this method to work, we assume that the mission requirements and goals are known, along with the target atmosphere and available support infrastructure (such as usable relay satellites, size and mass constraints etc.).

Phase I: Define Requirements.

Step 1 Compile mission goals and requirements and define their allowable ranges.

Step 2 Select an appropriate orbit and approach trajectory which best matches your available infrastructure.

Step 3 For the approach trajectory, choose a deceleration profile that best matches your mission requirements.

Phase II: Design a preliminary mission profile.

Step 4 Choose a ballistic coefficient range that will yield the desired deceleration profile (see chapter 7.2.1 starting on page 82).

Step 5 Define an appropriate atmosphere boundary (entry altitude) for the desired ballistic coefficient and approach trajectory (see chapter 3.5 starting on page 26).

Step 6 Calculate velocity and flight path angle at the atmosphere boundary for the previously defined approach trajectory and calculate an atmospheric entry trajectory (either numerically or analytically, as shown in chapters 7.2.1, starting on page 82 and 7.2.3, starting on page 101).

Step 7   Reiterate Step 2 through Step 6 until all critical parameters, most importantly entry trajectory and deceleration profile, best match your mission requirements and available infrastructure.

Phase III: Design a preliminary ballute spacecraft.

Step 8   Define the pod payload and draft a preliminary mass budget.

Step 9   Choose a ballute shape that meets your stability requirements (see chapter 4, starting on page 33).

Step 10   Choose an appropriate inflation gas and pressure range and estimate the ballute's own mass.

Step 11   Determine the desired inflation pressure range (see chapter 7.2.2, starting in page 96)

Step 12   With the preliminary mass budget of the instrument pod plus ballute and the drag coefficient of the desired ballute shape, the desired ballistic coefficient will yield the necessary drag effective face area of the ballute (see chapter 3.4, starting on page 22).

Step 13   With the drag effective area, calculate ballute dimensions and mass (see chapter 7.2.4, starting on page 116 and chapter 7.2.5, starting on page 124) and reiterate from Step 10.

Step 14   Run a preliminary entry trajectory calculation that yields preliminary mechanical and thermal loads.

Step 15   Choose a ballute material that meets your area density requirement and can withstand mechanical and thermal loads with sufficient margin (see chapter 6, starting on page 67).

Step 16   Calculate a new ballute mass and new ballute dimensions, to match the desired ballistic coefficient.

Step 17   Reiterate Step 10 through Step 16, until a consistent dataset is obtained.

Phase IV: Finalize mission design.

Step 18   Calculate the volume of the packed ballute (see chapter 8.4 starting on page 158) and design the TDS (see chapter 8 starting on 155).

Step 19   Choose a gas storage and inflation control system (see chapter 2.2, starting on page 4 and chapter 9.1, starting on page 169).

Step 20   Determine the gas storage and inflation control system mass and volume (see chapter 9, starting on page 169).

Step 21   Compare volumes and masses against the available infrastructure (transportation system).

Step 22   Reiterate Phases I through IV until all mission goals and requirements of the available infrastructure are met.

Note that this design method was developed while designing ballute missions to Mars, Earth and Venus (see following chapters) and has been used successfully even for quick look analyses. This is however only one general way of designing a ballute mission. In a real study, it might be practical to omit or rearrange some steps according to priorities.

10.2 The MIRIAM Spaceflight Test

For the evaluation of the entire ballute technology program, a representative test mission was designed and is outlined in chapter10.2.1 . The name of this test mission is MIRIAM, forming an acronym for. "Main Inflated Reentry Into the Atmosphere Mission test (for ARCHIMEDES)". This test is the direct continuation of the tests performed during the parabolic flight campaign in June 2005, as well as REGINA on REXUS 3 in April 2006.

For the REXUS 4 ballistic rocket campaign a MIRIAM experiment was designed and flown to test the release, inflation, separation and atmospheric entry of a hypersonic drag balloon.

10.2.1 General Mission and Spacecraft Overview

This chapter outlines the MIRIAM mission concept and describes details of the first attempt at flying a MIRIAM mission. Future MIRIAM missions can use this data as a reference.

For this test flight concept, MIRIAM is the mission as well as the ballute spacecraft's name. The space flight system composite consists of 3 major elements:

1. The Miriam ballute spacecraft which comprises an instrument pod and the helium-filled hypersonic drag balloon of 4m diameter and a total vehicle mass of around 5.6 kg .

2. The Service Module (SM) which contains the inflation system, structural box, release mechanism, SM telemetry and a live television subsystem. It also contains a set of cold gas thrusters. These thrusters need to pull the Service Module away from Miriam after inflation.

3. The Camera Module (CM) which stays on the rocket. It provides a platform for a camera subsystem and also provides the main release mechanism and structural interface for the Service Module / Miriam composite.

All three elements combined form the MIRIAM Flight System Stack. It needs to be mounted underneath the rocket's nose cone assembly, which has to be jettisoned to allow separation of the spacecraft. Alternatively, the rocket nose cone may double as the SM structure, if acceptable, which eliminates the need for nose cone separation and offers a potential for mass savings. For the REXUS-4 MIRIAM mission, this was however not possible.

A sketch of the actual flight system is shown in Fig. 10.1. It had a grand total of 39kg and stood over 1m tall. The overall diameter was 14 inches (355mm), a standard sounding rocket diameter. In this sketch, orange components belong to the rocket. Red components are avionics components, black are structural components and green are IGSS system deck components. Pink is the blossom container and clamp ring and blue represents observation equipment such as cameras and camera booms. Fig. 10.2 shows a picture of the service module structure and Fig. 10.3 shows the entire flight system stack during a shaker test at IABG Ottobrunn.

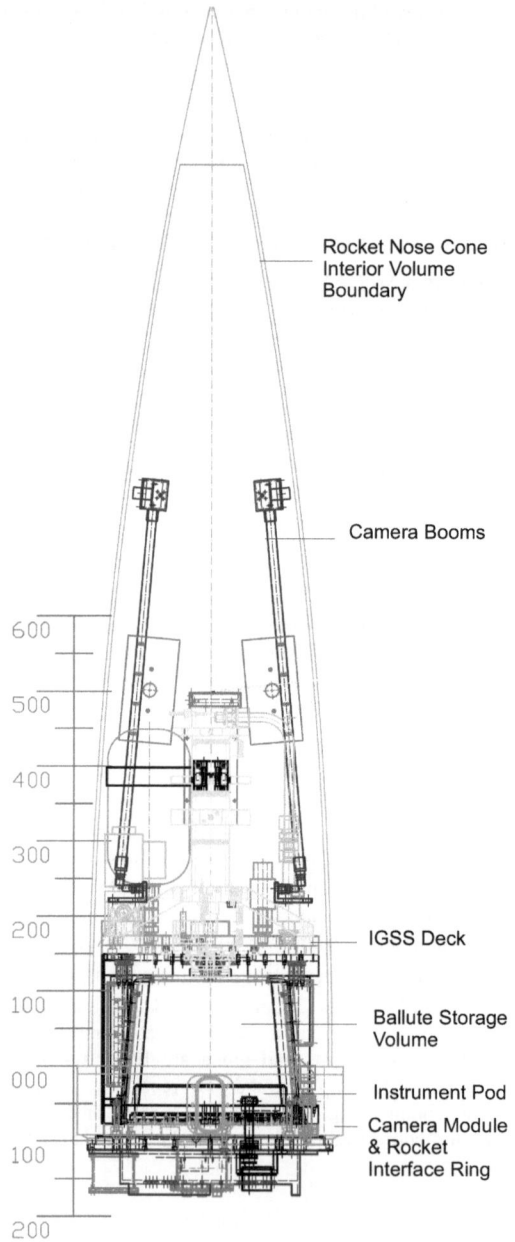

Fig. 10.1: Sketch of the MIRIAM Flight System Stack.

MRIAM's pod has to be instrumented for flight analysis purposes. On the actual vehicle, an FMI-Helsinki-provided ATMOS-B pressure sensor was installed behind the rear pod bulkhead, facing the ballute interior. The magnetometer for MIRIAM (MiriMag) is contributed by the IGEP institute and the company MAGSON of Berlin. Paired with an optical still image camera, it was to yield attitude information. The still image camera was a commercially available low resolution unit which could be integrated cheaply and easily and was sufficient for an occasional attitude fix, in combination with the other sensors. A suite of two different sets of accelerometers built by the ARCHIMEDES team and the universities in Iasi and Pitesti, Romania were to give deceleration and roll rate information.

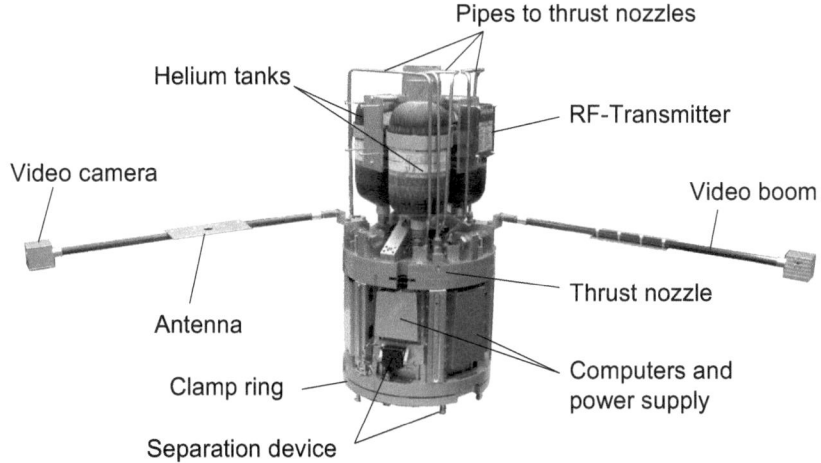

Fig. 10.2: The MIRIAM Service Module structure (actual flight model shown).

The MIRIAM flight sequence starts when either of two baro-switches inside the CM and SM register altitude. On the actual flight test the baro-switch prompted the two monitoring flight computers to notify each other of a successful climb and activated the flight control sequence on the PodCon computer.

The Service Module with the still packed Miriam Ballute has to move away from the payload section of the rocket (see also Fig. 10.24 on page 205). In a safe distance from the rocket, the Miriam deployment mechanism has to be activated. Cameras that are left in the Camera Module on the payload section can record the separation and deployment and may be recovered with the other payload.

Fig. 10.3: The MIRIAM flight system stack rigged for a vibration test at IABG space test centre Ottobrunn.

A pre-set time after the deployment of the Miriam ballute, during which the ballute is allowed to unfold under its own momentum and residual gas, the inflation control command sequence is initiated. After termination of this sequence, the balloon release mechanism opens. The Service Module is then pulled away from the Miriam ballute spacecraft, by means of gas thrusters, and the Miriam ballute is left to enter the atmosphere and begin its descent. The mission is terminates when reception of telemetry signals from Miriam and its Service Module ends. The full mission sequence for the actual MIRIAM-REXUS4 test is summarized in Table 10.1:

Event	Time [s]	Altitude [km]	Description
1	0	0.3	Lift-Off
2	4	1.4	Bunrout 1st Stage
3	35	36.5	Burnout 2nd Stage
4	55	68	Despin
5	59	73	Nose Cone Separation
6	64	87	MIRIAM P4MS (Prepare for MIRIAM Separation Subsequence)
6a	64		Camera Module Camera System on, Pod acknowledge command handover
6b	65		Service Module TV-camera boom deployment
6c	68		Service Module telemetry on, Pod starts instruments (data dump to internal memory)
6d	69		Service Module TV-transmission on
7	76	97	MIRIAM Separation (Service Module release from the Camera Module)
8	89	110	Miriam Clamp Ring release and Miriam Balloon deployment
9	99	119	Miriam Instrument Pod telemetry on
10	100	122	Inflation sequence start
11	220	180	Inflation sequence termination.
12a	222	179	Service Module Inflation Hose Clamp open
12b	224		Service Module Thruster System pressurization and Service Module pull away.
13	225	178	Miriam free flight mission starts
14	418	4.8	REXUS parachute deployment
15	420		Camera Module auto shut down in preparation for landing
16a	n/a		Miriam free flight mission terminates when the Miriam telemetry signal is lost.
16b	n/a		Camera Module recovery with the REXUS payload section
16c	n/a		Recovery of Miriam ballute spacecraft if retrievable
16d	n/a		Recovery of Service Module if retrievable

Table 10.1: MIRIAM mission event sequence. Altitudes are estimated from trajectory calculations done by the DLR / MORABA.

The Service Module telemetry and TV signals during the flight of MIRIAM were recorded, until the signal was lost (see also Fig. 10.24 and Fig. 10.25 on page 205).

In addition to transmitting live telemetry, the Miriam Instrument Pod and Service Module flight controllers were equipped with internal permanent solid state memories, to record certain flight parameters at higher resolution. To be able to retrieve the systems, both the Miriam Instrument Pod and Service Module were equipped with radio beacons.

Because MIRIAM is supposed to resemble a valid model of ARCHIMEDES, the ballute spacecraft was designed such that its aerodynamic and flight dynamic properties resembled the ones planned for ARCHIMEDES. The scale of the model is defined by the dimension relation of the ballute. Primary technical data can be found in Table 10.2.

Value	ARCHIMEDES	MIRIAM	Ratio M : A
Mass Instrument Pod [kg]	ca. 18.3	2.9	1:6.30
Mass Ballute [kg]	ca. 15.1	2.6	1:5.80
Mass Ballute S/C [kg]	ca. 33.2	5.4	1:6.10
Pod : Ballute [-]	ca. 1.21	1.12	1:1.08
Ballist. Coeff. [kg / m²]	ca. 0.54	0.59	1:0.92
Ballute Diameter [m]	ca. 10	ca. 4	1:2.50

Table 10.2: MIRIAM technical data compared to ARCHIMEDES.

10.2.2 Instrumentation and Methods to Achieve the Mission Goals

To evaluate the mission, several systems were present on the actual MIRIAM flight system stack, which had individual primary goals, but backed each other up as a secondary goal. These goals and instruments are briefly outlined hereafter.

Analysis of the deployment, inflation and separation behaviour.

Primary Systems

1. 2-Channel television subsystem, mounted on deployable booms on the service module.
2. Service Module telemetry of inflation system states.

Secondary Systems

1. Balloon interior pressure and temperature sensor readings through Miriam's pod telemetry.
2. Solid State imaging system on the camera module, reuse of REGINA hardware.
3. 1-Channel television camera on the camera module.
4. Postflight analysis of interior high-data volume solid state memory contents of the Miriam instrument pod (if retrievable).
5. Postflight analysis of interior high-data volume solid state memory contents of the Service Module (if retrievable).

Determination of the hypersonic flight behaviour of the given setup and subsequent trajectory calculation code validation:

Primary Systems

1. Miriam's pod telemetry readings of a 3-axis accelerometric measurement system, including 3-axis gyroscopes.
2. The AMS accelerometric measurement experiment.
3. The MiriMag magnetometric attitude sensor.

4. At least one image from the pod camera showing distinguishable features on either Earth or in the sky.

5. Solid state imaging system on the camera module for the photogrammetric analysis of Miriam's initial trajectory condition.

Secondary Systems

1. Balloon interior pressure and temperature sensor readings through Miriam's pod telemetry.

2. Postflight analysis of interior high-data volume solid state memory contents of the Miriam instrument pod (if retrievable).

Determination of the balloon material temperature and internal pressure during atmospheric entry and subsequent CFD and thermal model validation:

Primary Systems

1. One pod temperature sensor and 6 balloon skin temperature sensors.

2. The Atmos-B interior filling gas pressure and temperature sensor.

Secondary Systems

1. Postflight analysis of interior high-data volume solid state memory contents of the Miriam instrument pod (if retrievable).

10.2.3 Instrument Pod

The instrument pod of MIRIAM had an overall diameter of 235mm and a hexagonal primary structure milled directly out of high grade aircraft aluminium (see Fig. 10.4). The minimum effective fitting diameter used for stress analyses is that of a concentric circle, touching the O-ring seal around the hole circle (see also chapter 7.3.3 starting on page 135). Or in other words, the distance between the O-ring groove on two opposite straight sides. The effective minimum fitting diameter was thus 212mm. The instrument pod equipment layout is given in Fig. 10.5. The overall total mass was 2.9kg, of which the avionics suite and instruments took 700g (see Table 4.1). These are the essential basic values with which the ballute can be designed (see chapters 10.1 starting on page 179 and 7.2 starting on page 80).

Fig. 10.4: The MIRIAM Instrument Pod during a magnetic field simulation test at IABG Ottobrunn.

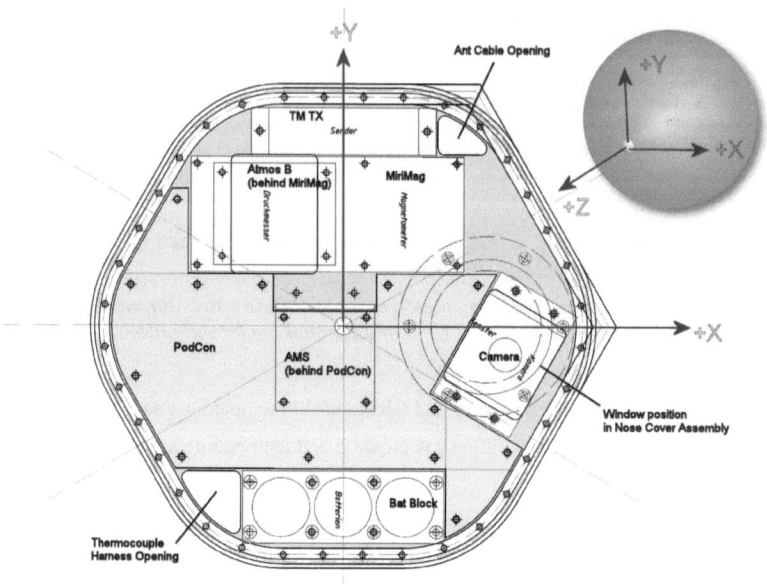

Fig. 10.5: MIRIAM instrument pod layout sketch, showing equipment location and coordinate system.

		Power Budget @ Voltage		Envelope Dimensions			Mass
		5	9,6	L	W or Diam	H	
Equipment	Model Designation	Current [mA]	Current [mA]	[mm]	[mm]	[mm]	[g]
Pressure Sensor	AtmosB	4,00	-	55,00	45,00	18,00	30,00
Accelerometer	AMS	30,00	-	85,00	80,00	17,00	20,00
Camera		60,00	-	20,00	28,00	17,00	50,00
MiriMag	MiriMag	-	83,00	124,00	51,00	35,00	123,00
Pod Controller	PodCon	70,00		150,00	20,00		150,00
Telemetry TX		-	833,33	68,00	43,00	17,00	120,00
Transmitter Antenna		-	-	-	1,00	16,00	30,00
Beacon (pwr internal)		-	-	25,00	30,00	17,00	30,00
Battery Cell 1	LSH26180	-	-	-	26,20	18,60	24,00
Battery Cell 2	LSH26180	-	-	-	26,20	18,60	24,00
Battery Cell 3	LSH26180	-	-	-	26,20	18,60	24,00
Battery Cell 4	LSH26180	-	-	-	26,20	18,60	24,00
Battery Cell 5	LSH26180	-	-	-	26,20	18,60	24,00
Battery Cell 6	LSH26180	-	-	-	26,20	18,60	24,00
Total		164,00	916,33				697,00

Table 10.3: List of MIRIAM Instrument Pod avionics, including power and mass budget.

10.2.4 Trajectory and Aerothermodynamic Analysis

Trajectory and entry analyses were performed according to the methods described in chapters 5 and 7.2, using both the CIRA and GRAM99 Earth atmosphere models [102]. For the analysis, a peak altitude of 180km and an entry angle of -70° in 100km were assumed. During the actual flight, 179km were reached. Results for the maximum deceleration and convective stagnation point heating rate are shown in Fig. 10.6. The plots show that the expected deceleration peak of 7.5g at 46km altitude far exceeds the value expected for the grazing ARCHIMEDES mission and would have most likely led to the destruction of the ballute. However, a trajectory of that type still delivers valuable flight data, that can be used to validate the mathematical models. According to these simulations, the maximum flow velocity is 1468 m/s corresponding to a flight Mach number of 4.8 and occurs at 68km altitude.

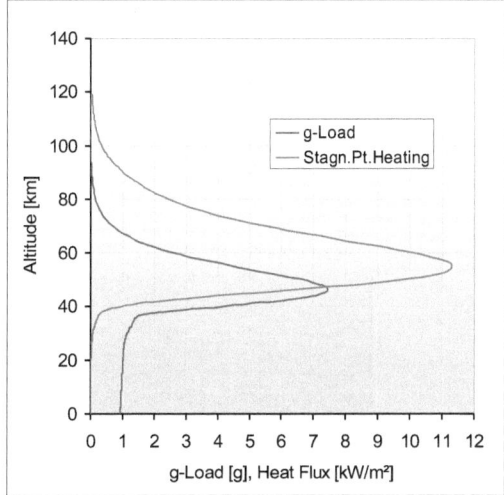

Fig. 10.6: Numerical calculation of the deceleration and convective stagnation point heating rate for MIRIAM.

Based on these results, CFD analyses with the CEVCATS-N code (see chapter 5.3, starting on page 50) and another code known as KEGSEC were performed [103]. The latter, developed by Prof. Ch. Mundt of the UBW, is based on solving the Euler equations in the shock front and then calculating the boundary layer conditions based on these results. The advantage is a drastic reduction in calculation time.

The ballute wall temperature resulting from both CFD analyses of the maximum convective heating point is shown in Fig. 10.7. Temperatures differ slightly, which was found to be an effect of the flow field grid, but are otherwise in good agreement with expectations and simpler estimates based on the method of Sutton & Graves [37]. A future grid study is therefore suggested, but was deemed unnecessary for the analysis of MIRIAM. The corresponding flow field temperature plot (in this case showing the KEGSEC results) is given in Fig. 10.8.

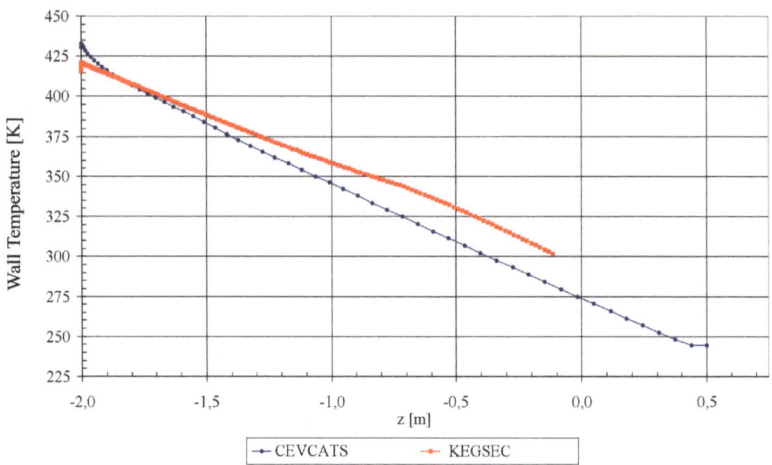

Fig. 10.7: Ballute wall temperature as calculated with two CFD analysis codes for the MIRIAM trajectory point of maximum heating rate.

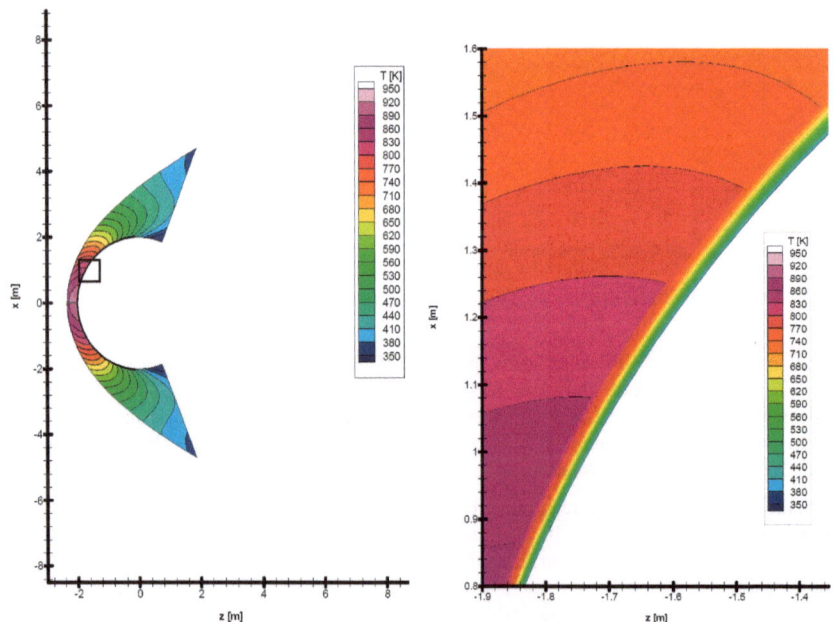

Fig. 10.8: CFD results for the point of maximum convective heating for the MIRIAM mission simulation obtained with the KEGSEC code.

The ballute was assumed to be filled with 55g of Helium for this analysis, leading to a nominal filling pressure of 10hPa. A peak interior pressure of 19hPa is expected from thermal analyses based on the method given in chapter 5.4 (starting on page 59). Because of the minute heat capacities and the large surfaces involved, maximum gas pressure occurs only shortly after peak heating.

According to this analysis, no temperatures may be expected for a MIRIAM mission that would prohibit the use of UPILEX-RN as a ballute envelope material. Local heating above the service temperature of the adhesive (see chapter 7.3.4.2 on page 138) however would likely weaken a seam or reinforcement patch above the point of structural failure, which is why the MIRIAM ballute is expected to disintegrate below an altitude of around 80km.

10.2.5 Ballute

The ballute was designed according to chapters 7.2.4 and 7.2.5 and manufactured together with the ballute team of the Mars Society Germany. To gradually improve ballute manufacturing techniques, the MIRIAM ballute was not optimized for performance as outlined in chapter 7.3.4 but designed limit manufacturing complexities. Nominal skin stress at 10 hPa inflation overpressure was found to be 39 MPa. At the point of maximum interior pressure the expected skin stress is 70 MPa. The highest fitting rim stress, which occurs at the point of maximum deceleration, amounts to 126 Mpa, with one additional reinforcement layer.

Fig. 10.9 shows a skin stress analysis (see chapters 7.2.4 and 7.2.5), without any safety factors applied. The worst case load was calculated with 65g of Helium, instead of the nominal 55g, to allow accidental over-pressurization, due to ICCS uncertainties.

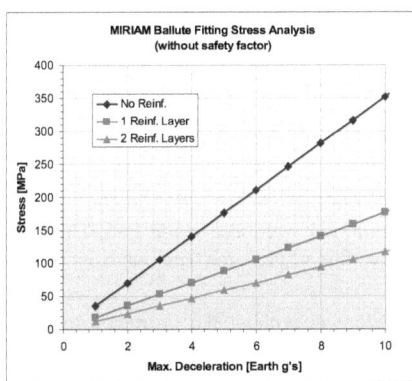 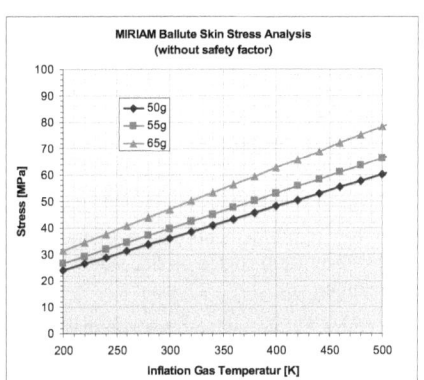

Fig. 10.9: MIRIAM ballute stress analysis with respect to deceleration and gas temperature, without safety factors.
Left: Instrument pod fitting rim stress for no reinforcement, as well as 1 and 2 reinforcement layers.
Right; Envelope skin stress for 50g, 55g and 65g of inflation gas.

The Miriam ballute was 4m in diameter and constructed entirely from UPILEX 25 RN [92]. The ballute had 32 longitudinal segments for each hemisphere and one continuous equatorial seam. All seams, reinforcements or equipment fixtures were bonded with the high-temperature high strength adhesive specially manufactured to MIRIAM mission specifications by Lohmann Tapes of Neuwied. Unfortunately, UPILEX-12.5RN remains as of now unavailable, although market introduction of the variant was repeatedly announced. UPILEX 25RN was therefore used as a tape substrate as well.

To limit the manufacturing complexity in this first space flight test as stated above, only 25 mm wide tape was manufactured. As Fig. 10.10 shows, the resulting adhesive shear stress is lower than for the 10-metre high performance design studied for ARCHIMEDES (see chapter 7.3.4.2 on page 138) and is thus acceptable. MIRIAM's smaller diameter results in a lower skin tension and therefore also lower seam load peaks, thereby mitigating the effects of the simpler seam design. Unfortunately, no shear modulus or shear limit is available for Lohmann's custom made adhesive layer, so typical values were again used for the estimate.

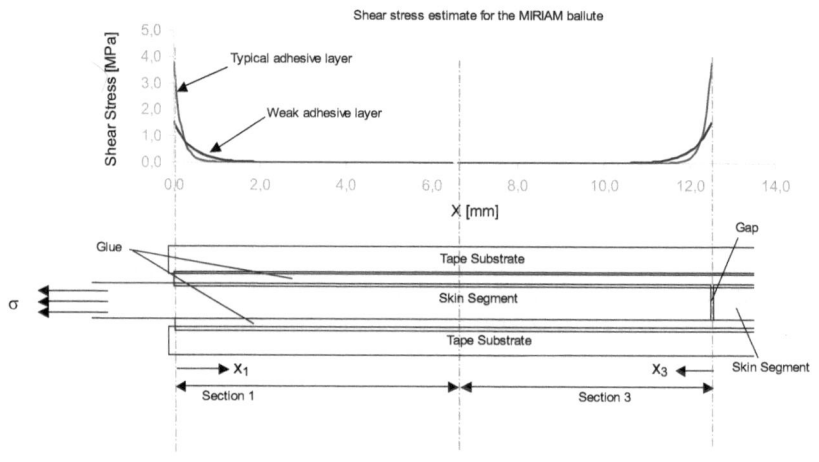

Fig. 10.10: Analytical shear stress analysis of MIRIAM's 4-m UPILEX-25RN ballute, with an interior overpressure of 10hPa and two 25mm wide UPILEX-25RN adhesive tapes. Only one side is shown, as the shear stress in the other side's adhesive layer must be the same because of the symmetry.

Polar cap reinforcements had a diameter of 600 mm and were also made from UPILEX-25RN. They were coated on one side with an adhesive film of the same type Lohmann used in the seaming tapes.

The entire ballute subsystem weighed a total of 2639 g, without the instrument pod but including all ballute instrumentation, an inflation hose and windsock assembly and the telemetry antenna. The MIRIAM ballute spacecraft is shown in Fig. 10.11.

After manufacturing was completed, the envelope instruments were tested and the ballute inflated for shape and size validation (see Fig. 10.11). A space simulation chamber test was performed at IBAG to validate the inflation control and gas storage system operation with the ballute under vacuum conditions and to validate gas tightness, which was confirmed by this test (see also Fig. 10.22 on page 203). After that the ballute was packed for flight.

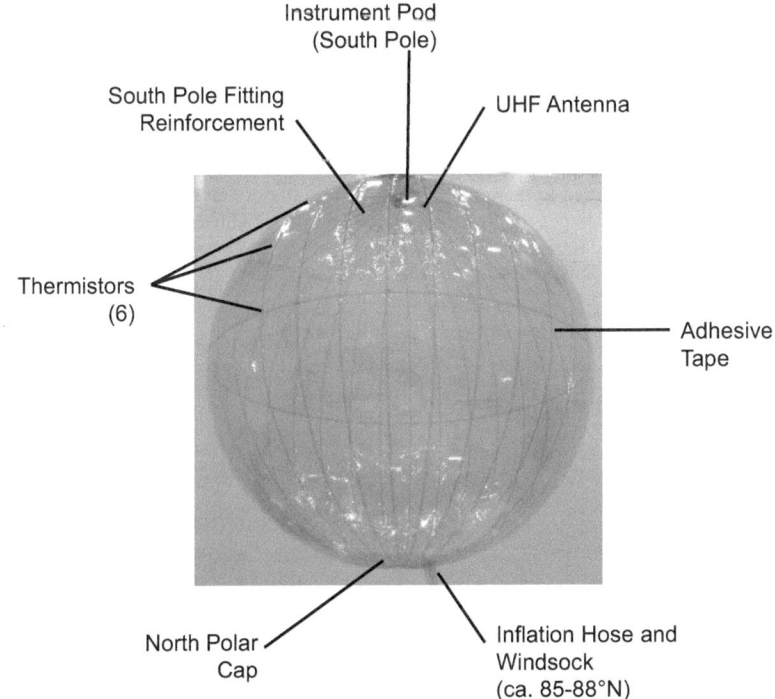

Instrument Pod
(South Pole)

South Pole Fitting
Reinforcement

UHF Antenna

Thermistors
(6)

Adhesive
Tape

North Polar
Cap

Inflation Hose and
Windsock
(ca. 85-88°N)

Fig. 10.11: The MIRIAM flight model ballute spacecraft during a pressurization test at IABG Ottobrunn.

Fig. 10.12: The ballute folded up in primary E-pleats and connected to the evacuation pump. (MSD)

The packaging procedure for MIRIAM was also developed together with the ballute team of the Mars Society Germany and carried out at a clean room facility at the IABG space test centre Ottobrunn. Packing was done using a custom designed frame made from standard "ITEM" aluminium bars. It had two rotating cylinders, supported by cross members. During primary e-folding on a large table, the ballute was evacuated through the instrument pod fitting hole with a commercial evacuation pump (see Fig. 10.12). The finished pre-folded band of E-pleats was then led over one cylinder to be rolled up on the other cylinder under tension. This cylinder served to store the primary band without drawing air and to hold it in place for secondary Z-folding. From there it was folded into an adjustable rectangular tray set to the desired width of the secondary Z-pleats.

Folding Frame   Storage Cylinder   Vacuum Pump

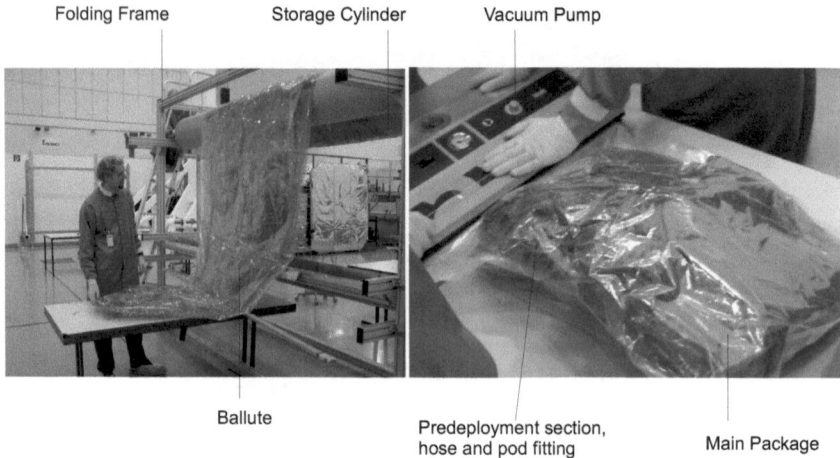

Ballute   Predeployment section,
                     hose and pod fitting   Main Package

Fig. 10.13: The packaging rig (left) and the MIRIAM ballute packed for flight (right). (MSD)

The long and narrow Z-pleated band was then divided into the pre-deployment section and the main section. The sections were densely packed and the south pole left sticking out for later installation of the instrument pod. The ballute wire harness was also left hanging out of the south pole fitting hole for later connection to the pod. The finished ballute package was then stored in an evacuated and sealed plastic bag for protection and to prevent the densely compressed package from falling apart prematurely. Fig. 10.13 shows the ballute packaging rig complete with tray, frame, cylinders and working surfaces at IABG Ottobrunn as well as the finished ballute package being sealed in an evacuated bag for storage.

10.2.6 Deployment and Inflation Subsystems

The Transportation and Deployment Subsystem (TDS) and Inflation Control and Gas Storage Subsystem (IGSS) to deploy and inflate the MIRIAM ballute spacecraft were two functional units. The IGSS was integrated into a separate deck that was directly attached to the rear TDS bulkhead.

Fig. 10.14: MIRIAM Service Module TDS primary structural frame and blossom, shown with a mass dummy of the IGSS system deck, rigged for modal tests at the Institute of Light Weight Structures, UBW [95].

A 3D sketch of MIRIAM's TDS assembly is given as an example in Fig. 8.1 on page 155. Details of the actual system can be seen in Fig. 10.15. It consists of a pyramidal and hexagonal structural frame that provides room for the blossom container in its centre and doubles as a support frame for avionics modules. The blossom is pushed forward by conical coil springs housed inside a flat, round cover that was attached to the blossom base on one side and to the TDS bulkhead on the other. The spring is conical so that it can be compressed to a height of one spring wire thickness and fits into the coil spring seat. The blossom's way of travel is limited by retention wires. The rear bulkhead provides a central opening for the hose clamp assembly. The complete TDS assembly can be seen rigged for modal tests on a vibration table at the UBW in Fig. 10.14.

Blossom Base   Coil Spring Assembly   Teflon Bearing

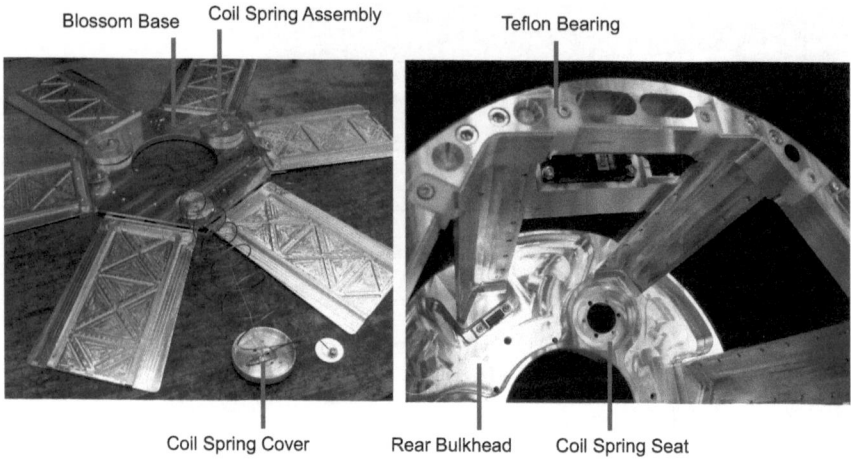

Coil Spring Cover   Rear Bulkhead   Coil Spring Seat

Fig. 10.15: Details of MIRIAM's blossom container (left) and TDS main structural frame (right).

10.2.6.1 IGSS System Design Overview

MIRIAM's IGSS is the first flight model implementation of the system described in chapter 9 and is depicted in Fig. 10.17. The entire inflation gas handling functionality is integrated into the inflation systems deck and is shown in Error: Reference source not found. This deck consists of a base plate supporting the plumbing, tanks and other, unrelated components. The Helium gas is stored in three CFRP pressure tanks at 200 bars. These tanks are carbon-fibre wound low cost tanks normally used to pressurize air guns for paint ball games. Because the tanks have a 300bar rated pressure, are TÜV-certified for use close to the human body, and tested up to 400bar, these components fulfil the MIRIAM requirements very well. In addition to this, they conveniently come with a 1.1 litre capacity, so three of these tanks can be integrated into the hexagonal structure to form a combined 3.3 litre storage volume. To reduce the piping between the tanks and the valves, most of the piping is integrated into the base plate of the inflation systems deck. This plate is divided horizontally into two parts and the gas pressure lines are milled directly into the lower half-plate (see Fig. 10.18). Sealing is done by conventional O-rings, with the O-ring groove milled into the top half-plate. Numerous tests on a shaker table and in a vacuum chamber including full flight cycles with full filling pressure have shown that leakage is not a problem [101]. The valves are connected directly to the channels in the base plate via special adapters with integrated gas channels. Shake tests confirmed that the valves seal tight at pressures above 150 bars, where the tank pressure pre-loads the valve anchor and thus prevents clattering when subjected to high g-load vibrations.

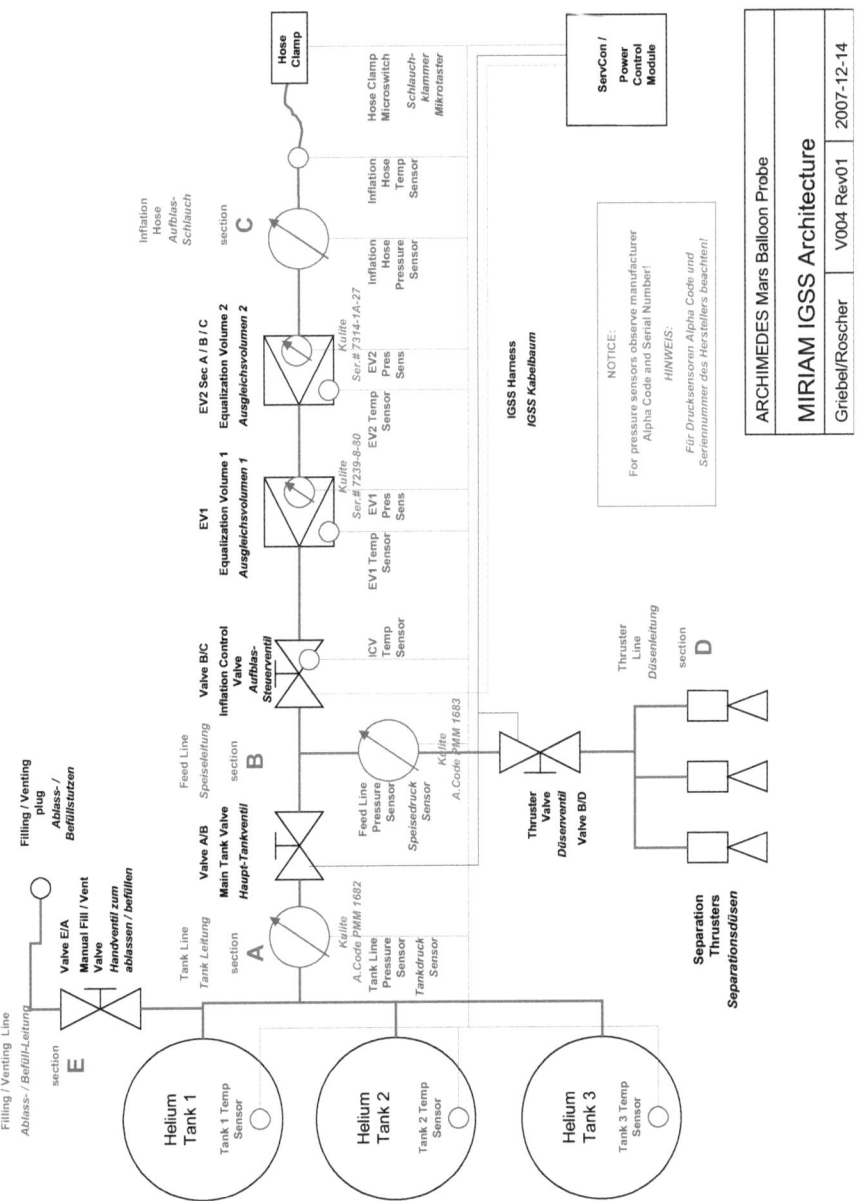

Fig. 10.16: MIRIAM's IGSS schematic diagram.

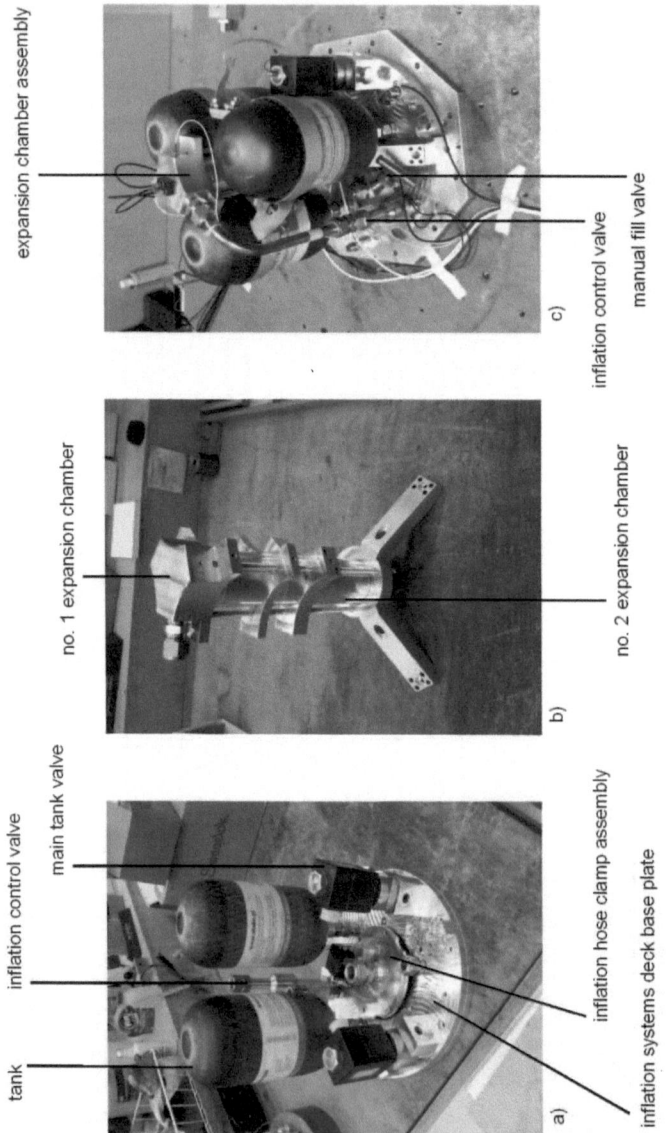

Fig. 10.17: The engineering model of MIRIAM's Inflation Deck Assembly: a) base plate with hose clamp, tanks and valves; b) expansion chamber assembly; c) complete inflation deck assembly rigged for vibration testing.

The only pipes necessary are those for connecting the channels with the thruster nozzles. One more is used to connect the main inflation control valve (ICV) to the central expansion chamber assembly. The two parts of the IGSS deck base can be seen in Fig. 10.18. Note the black O-ring seal in the left image.

Fig. 10.18: The two parts of the IGSS deck base plate, with the pressure feed system and tank system milled directly into the structure.

Because of severe volume restrictions on the small sounding rocket available for MIRIAM, the expansion volumes had to be very small. They have a combined volume of around 0.3 litres (approximately the volume of a soda can). If volumes are small, the ICV pulse width has to become short to achieve a large enough reduction in pressure. Therefore, a high-pressure high-speed solenoid valve which can open and close under 800 μs was used on MIRIAM. The unit was custom manufactured for MIRIAM by the company Hoerbiger of Vienna, Austria, who normally manufacture such valves for hydrogen combustion piston engine applications.

The inflation systems deck (or IGSS deck) was tested for launch vibrations, vacuum behaviour and operational limits.

10.2.6.2 The Inflation Control Command Sequence

On MIRIAM, a simple direct control ICCS procedure was implemented, without the complexity of a real-time feedback loop. The ICCS was pre-calculated using the simulation tool described in chapter 9.2.2 and tested successfully.

The flight control computer on MIRIAM's service module (referred to as Service Module Control Computer or, in short, ServCon) is based on an 8-bit ATMEL ATmega 2560 series micro controller. This CPU already offers a set of programmable 16-bit resolution pulse width modulation (PWM) channels. One such PWM channel is used to control the ICV through the Service Module's Power Control Module (PCM), which provides the ICV's high-power solenoid assembly with appropriate high-current pulses. However, floating point operations and multi tasking are not in the nature of such a small and simple CPU. In parallel to running down the ICCS, sensor readings and data handling tasks, such as telemetry preprocessing and data bus control functions have to be managed.

```
Inflation Control Command Sequence No. 008
; Intention: Vacuum Chamber Test, Flight
; Date:       25/03/08
; Author:     A.Barth / H.Griebel

; Cmd#        Pulses        PlsWdth        Hz
1             200           1250           20
2             100           3250           20
3             100           3630           20
4             100           3380           20
5             100           4130           20
6             100           4500           20
7             100           5000           20
8             100           5500           20
9             100           6250           20
10            100           7000           20
11            100           8000           20
12            100           9000           20
13            100           11000          20
14            200           13000          20
15            800           1250           20
done
```

Table 10.4: Inflation Control Command Sequence for MIRIAM

To prevent the necessity of a separate inflation control computer, certain restrictions applied to MIRIAM's ICCS list entries which are summarized as follows:

- Only integer values can be used.
- The maximum number of commands is 100.
- The control frequency has to be a multiple of 10 and smaller 100.
- The ICCS list file has to be formatted to precise rules.

The ICCS list was uploaded to the ServCon as an ASCII file through MIRIAM's ground support equipment by an UART link connected to a PC with an RS232 interface. The formatting rules of the ICCS file were:

- Empty lines and lines beginning with a semicolon are ignored.
- The first line which is not empty and not a comment must be a precisely coded header with a version number.
- All other lines contain command entries, one command per line, command parameter separated by TAB or spaces. Command parameters are: Command Number, Number of Pulses, Pulse width in µs, Frequency in Hz.
- Commands with Frequency 0 Hz are interpreted as uninterrupted open, where pulse width is then interpreted as the opening time in seconds.

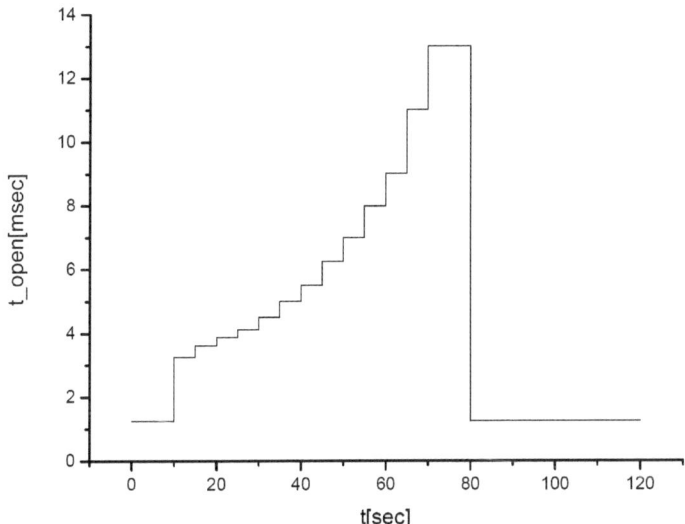

Fig. 10.19: ICV Opening times for the MIRIAM flight ICCS [101].

The ServCon had a built-in plausibility check that rejected any uploaded sequence containing a command with a pulse width greater than 1/Frequency (meaning that the pulses would be longer than the possible time for an open/close cycle at a given frequency). Files with command list entries that had conflicting or not ascending command numbers, an incorrect header or a version number equal to or smaller than the one already stored in the computer's command memory were also rejected. This was a precautionary measure to prevent accidental copy and paste errors or the upload of a wrong or old file that could have led to a malfunctioning of the system. It also forced the operator to check the command sequence list stored in the computer.

The ICCS found to work best (which was used for all major tests and later the flight) is given in Table 10.4 and Fig. 10.19 [101]. The goal was to build up gas pressure (and throughput) in two steps, to prevent the partly deployed ballute from rupturing and to prevent too steep of an increase in pressure, which might have pushed the inflation hose off the hose clamp assembly. It was then to be kept at a constant level, to ensure a sufficient rate of inflation. To approach the design load of the ballute skin gradually, the pressure was reduced with the final inflation command, so that tension in the ballute skin would build up gradually, rather than abruptly. The simulated ballute pressure differential is given in Fig. 10.20, while simulated system and ballute pressures are given in Fig. 10.21. The large areas in these graphs are pressure fluctuations too small to be recognizable at this resolution. Since this is an ideal gas simulation assuming steady state conditions at each time step, real world fluctuations are not expected to be nearly as pronounced. Besides, deflector plates inside the expansion chamber assembly are also not modelled.

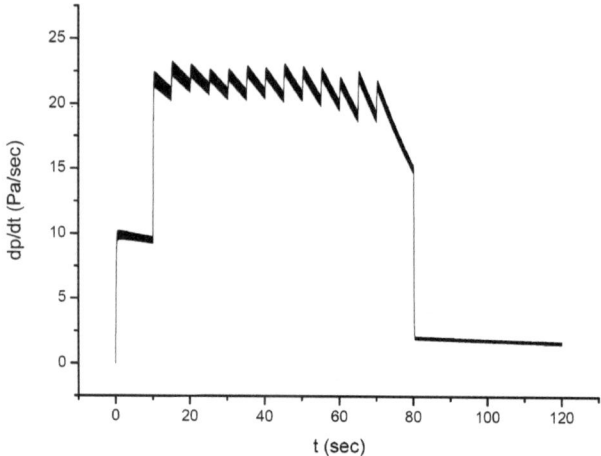

Fig. 10.20: Windsock-Ballute pressure differential for MIRIAM's flight ICCS [101].

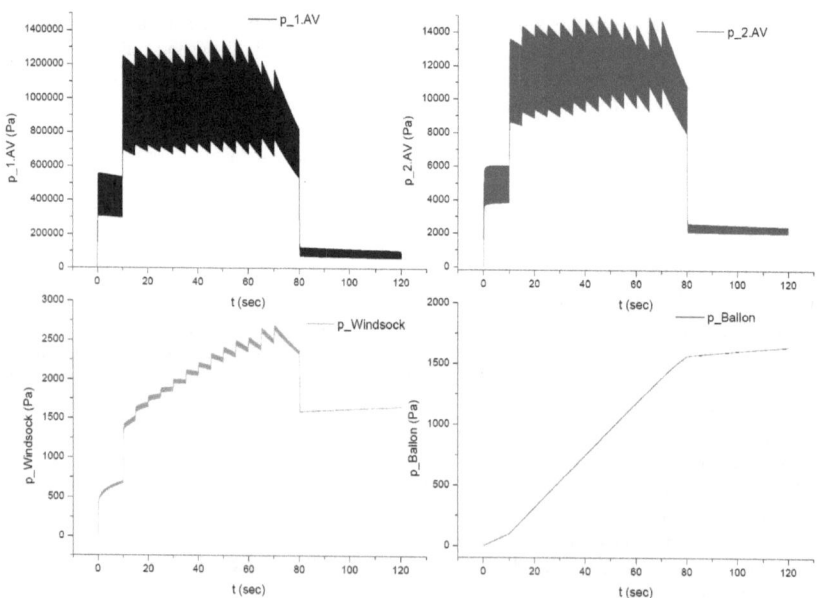

Fig. 10.21: MIRIAM's flight ICCS simulation results for the number 1 and number 2 expansion volume pressures (AV1 and AV2), windsock and ballute (balloon) pressure [101].

This inflation control command sequence was then uploaded to the engineering model of the flight system and tested with a geometric copy of the windsock in the lab, as well as a vacuum chamber. During the latter test, the vacuum chamber's volume acted as the windsock and ballute's combined volume, as the chamber itself is too small to accommodate MIRIAM's 4 metre ballute. In comparative simulations for lab tests, the surrounding atmosphere was modelled, assuming an ambient pressure of exactly 1 bar. In comparative simulations for the vacuum chamber tests, the windsock and ballute volume were replaced by that of the chamber. The recorded tank pressure was then compared to the simulated tank pressure (see Fig. 10.23). The comparison of tank pressure is a good method of evaluating the validity of the model and the ICCS, because the gas depletion rate within the tank causes a mass flux which, because of the continuity condition give in equation (188), has to remain constant throughout the system.

Fig. 10.22: Inflation test in the space simulation chamber at IABG Ottobrunn. a) inflated ballute as seen from one of the camera booms. b) inflated ballute with service module seen from the lower observation camera c) after chamber pressurization. (MSD / IABG)

In Fig. 10.23, the ICCS starts at 0 seconds and ends at 120 seconds. The experimental measurement starts at a tank pressure less than a bar shy of 180 bars. The short peak at around 10 seconds is an artefact. Significant, however, is the slight delay of the actual measured pressure drop, which is due to real gas effects and friction, which has been neglected for this simulation. Also noteworthy is the continuing slight drop in pressure after the ICCS has been completed and all valves have been shut. The reason for this is that right at the time where the valve shuts, pressures are not fully equalized throughout the systems and the valves are not fully tight at the instant they close. Therefore, the pressure takes a couple of seconds to stabilize.

The IGSS was eventually tested, together with the ballute, under simulated space conditions in IABG's space testing facility. Fig. 10.22 shows pictures taken during and after the test. Note the crossmember supporting part of the ballute weight, while a carefully balanced counterweight supported the rest of the ballute spacecraft. This prevented the inflation hose and windsock from collapsing and to further prevent the ballute spacecraft to sag too far and spread across the floor and equipment.

The tests show that the system was tight and working as intended. They also confirmed that mass throughput into the ballute can be predicted reasonably well with this simple simulation. Visual observation of the windsock also confirmed a strong and firm windsock pressurization, a sufficiently wide feed-through hole into the ballute envelope and no critical condition (such as severe vibrations, rupture through overpressure or dislodging of the inflation hose on the hose clam p assembly. Therefore, the simulation model along with the ICCS were considered qualified for analysis purposes and the flight.

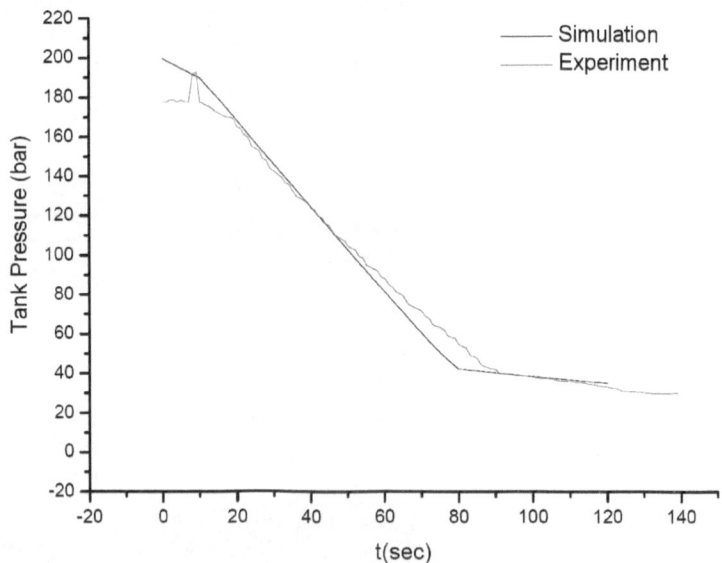

Fig. 10.23: Comparison between simulated and experimental tank pressure during a vacuum chamber ICCS test.

10.2.7 Flight Report and Conclusion

The MIRIAM flight system stack was delivered to ESRANGE on Thursday, October 16th, 2008. On October 20th a flight software update was installed on the Camera Module Control Computer (CamCon). During pre-flight tests, all flight system elements of MIRIAM performed nominally. The spacecraft was commissioned for flight on the evening on October 20th, 2008. The batteries were fully charged on October 21st, the instrument pod of the MIRIAM entry vehicle (ballute) was armed and the service module mounted on top of the REXUS-4 payload section. A preliminary pressurization of the service module tanks was performed on the evening of October 21st with Helium supplied by ESRANGE.

In the early morning of October 22nd, 2008, the service module was fully pressurized to 200bars, the flight-mode key installed, thus arming the service module and the nose cone was mated to the payload section of REXUS-4.

During test count-down early on October 22nd, MIRIAM's instrument pod failed to report flight-standby mode. This was an abnormal behaviour and not observed during tests. Because all other computers reported nominal conditions and other subsystems and experiments on REXUS-4 appeared ready for launch, the decision was made not to address this anomaly. This decision was further supported by the unfavourable weather forecast for the following week, which made a postponed launch of REXUS-4 during the allotted launch window unlikely.

The countdown was resumed and completed without incident. Upon ascent, both baro-switches on board MIRIAM correctly registered altitude and all computers, except the instrument pod, successfully reported switching into flight mode. At t +64 seconds the P4MS (prepare for Miriam Separation) subsequence was started and concluded successfully several seconds later, with the successful deployment of the camera booms on the service module. The TV and telemetry subsystems were started on time. The instrument pod did not report execution of P4MS. While it did wake up from hibernation mode, the instrument pod fell back to sleep repeatedly for reasons which are still a subject of investigation.

At 76 seconds, the main interlock actuators that secured the flight system during launch were turned to release. However, MIRIAM's main interlock bolt number one jammed upon release. The reason for this behaviour remains currently unknown and is a matter of investigation. This behaviour was not observed during pre-flight tests.

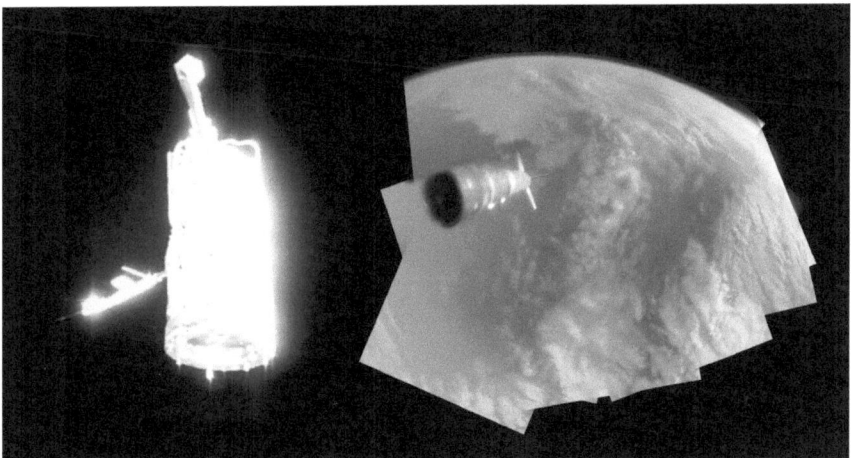

Fig. 10.24: The MIRIAM Service Module, as seen from the rocket payload section (left) and the payload section, as seen from the service module (right).

As a result, the MIRIAM Service Module stayed on the rocket, which subsequently led to the balloon's clamp ring to jam as well. 13 seconds later the clamp-ring interlock actuators were rotated to release. This caused the flight system to finally release from the rocket. This was, however, too late in the separation sequence and with the deployment coil springs almost fully extended, separation was far too slow. Pictures of the service module and the rocket payload section can be seen drifting away from each other in Fig. 10.24

At 100 seconds, the inflation control command sequence was started and the inflation systems-deck started to operate nominally. With the clamp-ring still partially attached the inflation sequence continued. This led to a rapid increase of pressure inside the inflation hose, the hose clamp assembly and the expansion chamber assembly and also increased the pressure against the jammed clamp ring and the hose attachment clamp.

The rocket payload was then separated from the rocket motor, thus gaining forward impulse. The rocket payload section missed the main bus but impacted the left camera case on top of the left camera boom at roughly 115 seconds.

Unintentionally positive however, the collision caused the clamp ring to shake loose, causing a rapid deployment of the partly pressurized balloon. Because the balloon was only attached to the inflation systems deck at the inflation hose clamp assembly, overpressure inside the hose clamp assembly immediately pushed the balloon off the hose clamp. Therefore the balloon was immediately set free, but with little over 10% of its intended amount of filling gas. A picture of the ballute in space is Fig. 10.25.

All other subsystems functioned nominally and behaved as expected from tests.

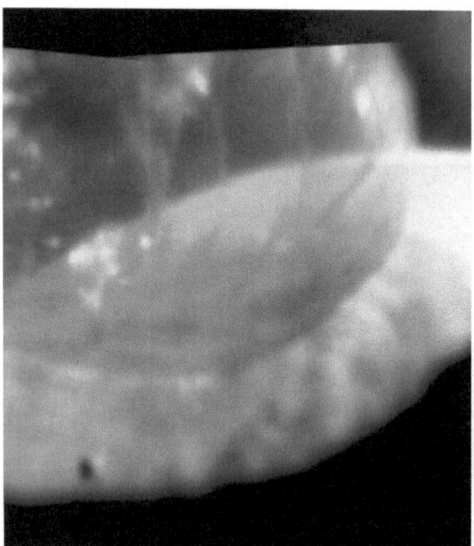

Fig. 10.25: The MIRIAM ballute spacecraft during deployment.

So far, only the Camera Module which stayed on the payload section could be recovered. The beacon signal of the Service Module was picked up by the helicopter search crew. On the morning of October 23rd, a helicopter mission in search of the Service Module was flown. While the beacon signal was picked up and could be tracked, only the rocket's upper stage motor was found. This is plausible, since both bodies enter the atmosphere on a ballistic trajectory, with similar aerodynamic properties. The Service Module itself however could not be spotted. Due to the very deep impact of the upper stage motor, it is believed that the Service Module dug a crater much deeper than the module is long and therefore remains hidden.

The following systems performed nominally:

1. The inflation systems deck.
2. The inflation control system.
3. The flight computers on the Service Module and on the Camera Module.
4. The telemetry subsystem.
5. The television subsystems.
6. The recovery beacon.
7. The boom deployment mechanism.
8. The deployment actuator systems.
9. All other electrical subsystems.
10. All camera systems, with the exception for the Number 2 solid state video recorder.
11. All ground and support hardware.
12. The balloon.

The following systems have not performed nominally:

1. The instrument pod control computer failed to report flight standby mode.
2. The main interlock number one jammed when rotating to release, which also caused the clamp ring interlock to hang.
3. The ServCon halted telemetry processing during ICCS execution.

None of the off-nominal behaviours have ever been observed during ground tests, so a combination of flight conditions that could not be reproduced on the ground is expected to be the likely cause.

The ballute was deployed and released and the mission is considered a partial success. Unfortunately, the deployment was unintentional and uncontrolled, with the consequence that the balloon inflation was not complete.

10.3 The ARCHIMEDES Mars Balloon Probe

ARCHIMEDES is a hypersonic drag ballute concept designed to probe the atmosphere of planet Mars. The project is currently under study by the Mars Society Germany and the Universität der Bundeswehr München and also supported by several other research institutions and industrial companies. The probe is a suggested payload to AMSAT's P5-A Mars satellite project, intended to be released from the spacecraft when in orbit around the planet. Launch of the P5-A is planned as a piggyback payload on an Ariane V rocket, as it is standard practice for spacecraft of the German AMSAT section.

Since many technical details of this project are already explained in respective chapters, this chapter is intended to give a more general overview of the concept. An artist's rendition of the ARCHIMEDES ballute approaching the Martian atmosphere is shown in Fig. 10.26.

10.3.1 Scientific Mission and Payload Instruments

The scientific scope of project ARCHIMEDES involves in-situ measurements in the Martian atmosphere, magnetic environment and surface throughout almost the entire altitude range reaching from outer space to ground [1].

Fig. 10.26: Artist's impression of the ARCHIMEDES ballute, approaching Mars [95].

Another important goal of the project is to demonstrate and qualify the ballute technology for entry into planetary atmospheres at high velocities on a representative mission.

The suggested payload suite is reflected in the mission name, which forms an abbreviation for "Areal reconnaissance Robot Carrying High-resolution Imaging, a Magnetometer Experiment and Direct Environmental Sensors". Hence, the primary payload constitutes a high resolution camera suggested by the DLR centre for planetary exploration Berlin, a magnetometer experiment studied jointly by the IGEP institute of the technical university of Braunschweig and the private company MAGSON of Berlin and the so called ATMOS-B weather sensor suite by the Finnish Meteorological Institute (FMI) of Helsinki. Additionally a pyrolytic compression wave temperature experiment by the IRS institute of the Technical University of Stuttgart and a high sensitivity accelerometer built jointly by the Technical Universities of Iasi and Pitesti in Romania are foreseen to ride in the nose cover assembly, which will be jettisoned after the transition to subsonic speeds.

To augment the science return of the mission, a low weight low power radar altimeter is currently under development for ARCHIMEDES, which will give altitude accuracy between 1 and 10 meters for a range of up to 100km [104].

Table 10.5 gives an overview of ARCHIMEDES' active scientific sensors. Note that the altimeter is not a scientific instrument by itself and therefore included in the housekeeping data.

Science Payload: Instrument Properties						
	Camera Sensor	Magnetometer	ATMOSB	COMPARE	AMS	Housekeeping*
Sensor, Sign. Conditioning	0.40 kg	0.15 kg	0.20 kg	1.15 kg	0.50 kg	0.00 kg
Houskeeping Data	64 Bit	64 Bit	64 Bit	64 Bit	64 Bit	0.00 Bit
Data / Sample	30,000,000 Bit	60 Bit	1,600 Bit	24 Bit	64 Bit	581 Bit
Data Compression Factor	10 -	3 -	3 -	3 -	3 -	3 -
Sampling Rate Continous		1 Hz	4 1 / min	50 Hz	50 Hz	0.10 Hz
Sampling Rate Non-Continous			30 1 / h		0	0.00
Sampling Rate SKY-MODE	2 1/h	1.00 Hz	0 1/h	0 Hz	10 Hz	0.10 Hz
Sampling Rate HY-MODE	12 1/h	1.00 Hz	0 1/h	50 Hz	10 Hz	1.00 Hz
Sampling Rate LOW-MODE	3 1/h	1.00 Hz	30 1/h	0 Hz	10 Hz	0.10 Hz
Power Up Time	1 s		0 s	0 s	0 s	0.00 s
Read-Out Time / Sample	9 s		15 s	0 s	21,600 s	0.00 s
Operating Power	1.00 W	0.50 W	0.50 W	1.00 W	0.50 W	0.00 W
Standby Power	0.00 W	0.00 W	0.00 W	0.00 W	0.00 W	0.00 W
* Housekeeping is treated as a science instrument for data budgets. Power consumption and mass budgeted with individual units.						

Table 10.5: Overview of ARCHIMEDES' proposed scientific instruments and their technical data.

Three instrument requirements which directly impact mission design are the following: firstly, in order to function as intended, the camera has to point in different directions, preferably scanning the surroundings with a down-looking and a side-looking camera head. The camera also needs daylight to take pictures. Secondly the ATMOS-B suite can only function below a certain airflow velocity and must therefore be protected from the hypersonic environment during entry. And thirdly would the AMS, Magnetometer and COMPARE experiments benefit greatly from repeated atmospheric passes, because they could collect more data and also collect it repeatedly.

10.3.2 Mars Mission System Elements

The mission has three major flight system elements: the AMSAT P5-A Mars orbiting satellite, the joint AMSAT-Archimedes Propulsion System JPS and the ballute system ARCHIMEDES (see Fig. 10.27).

ARCHIMEDES and its TDS are mounted inside the central structural core of the JPS. The JPS houses the cruise propellant tanks, the ARCHIMEDES gas storage and inflation system (IGSS) and the 400-N cruise engine. The P5-A satellite contains the solar arrays and power system, a 2-m high gain dish antenna (HGA) and an array of low gain omni antennas (LGA). It also contains the reaction control system (RCS), attitude control system and navigation and telecommunication subsystems.

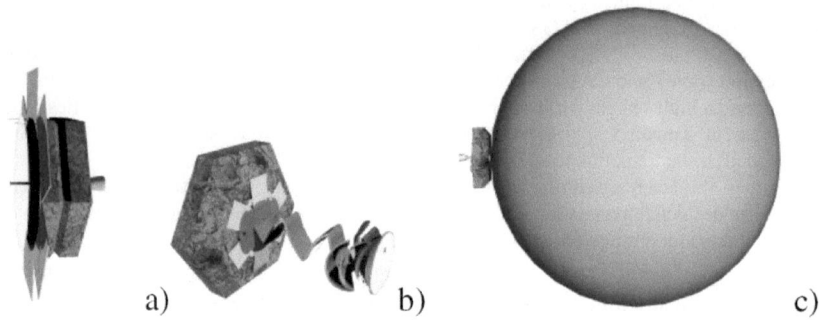

Fig. 10.27: a) The AMSAT P5/A in cruise configuration. b) The ARCHIMEDES vehicle being deployed from the Joint Propulsion System Module (JPS). c) ARCHIMEDES fully inflated with the JPS module still attached to the North Pole.

Total projected Earth orbit injection mass is 650kg, of which 110 kg are the orbiter. ARCHIMEDES accounts for 66 kg, including the 34 kg ballute spacecraft, plus 32 kg inflation and transportation systems. A 20% mass growth allowance is added to this budget, to account for design uncertainties, so that the total mass reserved for ARCHIMEDES in the combined mission study is 80 kg.

10.3.2.1 The ARCHIMEDES Ballute Spacecraft

The ARCHIMEDES ballute is designed to be a Helium-inflated sphere of 10m diameter made of high temperature resistant thin-film material, carrying a 4.5 kg instrument pod with a nose cover assembly (NCA) of 10kg clamped to one of its poles. The total atmospheric entry mass of the ARCHIMEDES vehicle will be approximately 34 kg.

The magnetometer experiment, the imaging system, the weather experiment and the altitude radar are installed in the main instrument pod. Especially the ATMOS-B sensor suite needs to be protected from the hypersonic flow environment by the nose cover assembly.

The NCA houses the accelerometer package and the pyrolytic flow field experiment COMPARE, along with additional batteries. The NCA can be jettisoned to expose the Instrument pod to the environment and to lighten the craft of some unused mass that would otherwise only shorten the descent time.

The ballute spacecraft is shown in an artist's rendition of the moment the nose cover assembly is discarded in Fig. 10.28.

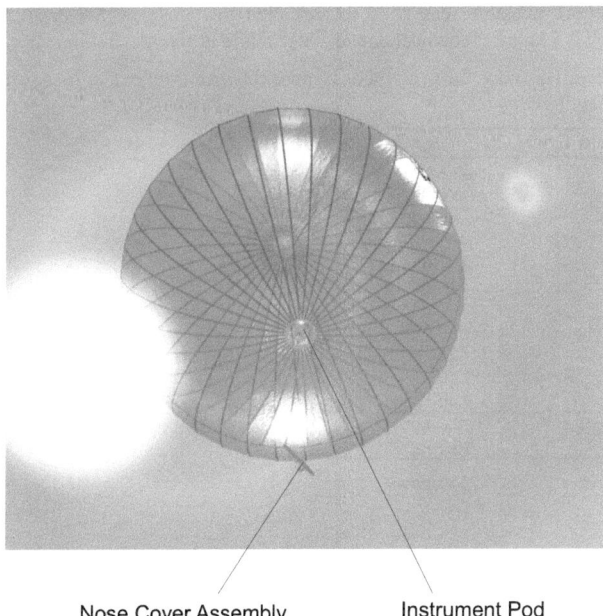

Nose Cover Assembly   Instrument Pod

Fig. 10.28: Artist's impression of the ARCHIMEDES ballute spacecraft, descending through the Martian afternoon sky. Note the nose cover assembly dropping away [95].

Power is provided by two primary battery packs, one each on the main pod structure and the NCA. They were designed to support even a very shallow, long lasting mission (see chapter 10.3.3). With this mission simulation data, the mission requirements obtained from [1] and scientific payload instruments from Table 10.5, an instrument pod can be designed. The mass budget of pod and NCA is given in Table 10.6 and Table 10.7, respectively.

Instrument Pod Mass Budget (kg)	
Structure	1.2
Battery	0.6
Electronics	0.7
e-box Casing	0.5
Antenna	0.2
Thermal	0.3
Add. Cabling	0.4
Science Sensors	1.0
Total Pod	**4.8**
Margin Pod	**0.5**
Total Pod with Margin	**5.3**

Table 10.6: ARCHIMEDES Instrument Pod mass budget.

Nose Cover Assembly Mass (kg)	
Structure	4.6
Thermal	0.3
Additional Cabeling	0.4
Electronics	0.1
Science Sensors	1.7
Batteries	1.7
(Margin of NCA in Non-Pod Mass)	
Total	**8.7**

Table 10.7: ARCHIMEDES Nose Cover Assembly mass budget.

With the instrument pod mass, a ballute can then be designed that matches the desired ballistic coefficient of around 0.5 kg/m². Technical data of this ballute is given in Table 10.8.

The resulting ballute spacecraft is now fully defined. A mass budget (without nose cover assembly) is given in Table 10.9.

Ballute Body	
Gasconst Filling Gas (He)	2077 J/kgK
Filling Pressure (nominal)	8 hPa
Filling Gas Density	0.0016 kg/m^3
Envelope Volume	524 m^3
Envelope Diameter	10.00 m
Envelope Surface Area	314 m^2
Envelope Film Aerea Density	37 g/m^2
Envelope Skin Material Composite	UPILEX-RN / PBO
Skin Thickness	25 µm
Hull Production Factor	1.40 -
Weight of Seams, Fittings etc.	4.65 kg
Min. Weight of Filling Gas	**0.8 kg**
Min. Weight of Ballute	**16.3 kg**

Table 10.8: Design parameters of the ARCHIMEDES ballute.

Ballute Spacecraft Mass (kg)	
Pod (with 10% Margin)	5.3
Ballute Envelope	16.3
Inflation Gas	0.8
Total	**22.4**

Table 10.9: ARCHIMEDES ballute spacecraft mass budget (without NCA).

10.3.2.2 The Joint Propulsion System (ARCHIMEDES' Service Spacecraft)

The JPS is part of the AMSAT P5-A / ARCHIMEDES spacecraft composite. It provides propulsion during cruise and doubles as a service spacecraft for ARCHIMEDES, once the desired orbit has been reached (see also chapter 10.3.3) and cruise propulsion is no longer needed. It consists of a flat, hexagonal structural box with six compartments. These house the UDMH/NTO liquid fuel propulsion system. Two oxidizer and two fuel tanks feed an Astrium manufactured 400N engine. The fourth and fifth compartment are used for the tank pressurization system, avionics and the IGSS.

Once separated from the orbiter, the JPS engine can be used for manoeuvring the JPS/ARCHIMEDES composite onto the right approach trajectory for atmospheric entry. While a spin stabilization was chosen for the JPS, a 3-axis stabilization option was also studied and might still present a favourable alternative.

The TDS resides inside the central structural core of the JPS. In this position, the ballute spacecraft receives the best possible radiation protection during cruise, because it is surrounded by mass on either side, as well as a lot of propellant. The packed ballute spacecraft is held by a blossom container which opens, like on MIRIAM, through a spring-loaded mechanism, towards the side opposite to that of the rocket engine. The clamp ring is attached to the top of the spacecraft. The side panels of the blossom container rest against the inner wall of the spacecraft and are kept on distance by Teflon® gliders to prevent jamming. As soon as the clamp ring is released, the springs will push the container forward, until retention wires hold it back. The impulse of the instrument pod will carry the now free package further, unfolding the pre-deployment sections of the ballute with a speed of approximately 1 m/s (see Fig. 10.29).

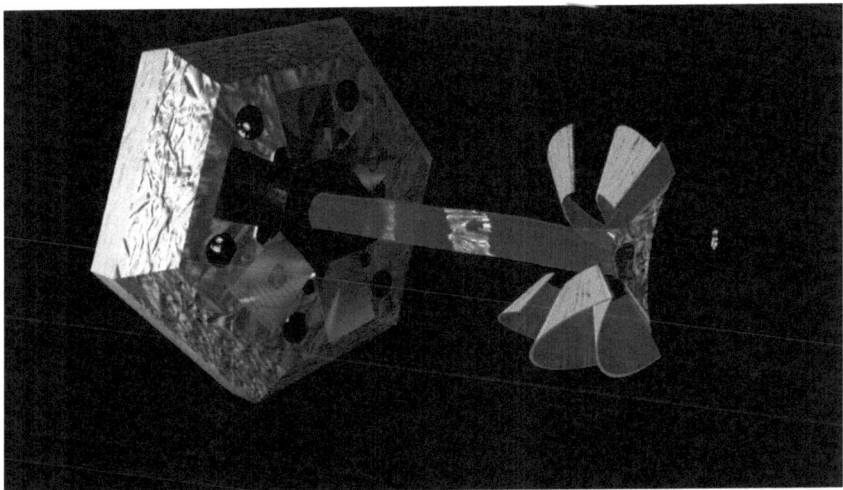

Fig. 10.29: Artist's impression of the joint propulsion system shortly after deploying the ballute spacecraft, which expands as it travels forward [95].

A summary of the JPS mass budget is given in Table 10.10. A detailed mass and power budget of JPS subsystems including the TDS, IGSS, propulsion system and attitude control elements. as well as avionics and telecommunication subsystems, was omitted here to maintain the focus on the ballute spacecraft.

Joint Propulsion System Module Mass Budget	
Total Budgeted Cruise dV	3,000.0 m/s
Specific Impulse	309.8 s
GTO Departure Wet Mass	549.2 kg w/o margin
Pres. Tank Mass	3.0 kg
Total Prop Tank Mass	17.6 kg
Engine Mass	3.0 kg
Propulsion Structure	23.7 kg
Inflationsystem and Tank	18.0 kg
Total Propulsion Dry Mass	**65.3 kg**
Cruise Propellant	344.5 kg
ARCHIMEDES DeOrbit Propellant	8.4 kg
Mixture Ratio	4.5 Ox / Fuel
Total Propellant Mass	**352.9 kg**
Total JPS Mass	**418.2 kg**

Table 10.10: Joint Propulsion System mass budget without margin. Cruise dV budget, engine and propellant data supplied by AMSAT.

10.3.3 Mission Design

The JPS provides propulsion for all manoeuvres up to the deployment of ARCHIMEDES. The JPS will also function as a self-contained spacecraft once detached from the satellite, running on internal primary batteries for all orbit manoeuvres required to bring ARCHIMEDES onto its correct approach trajectory and the subsequent deployment and inflation of the ballute spacecraft. Once this has been accomplished, the JPS uses cold gas thrusters to separate from ARCHIMEDES and thus discards itself.

When detached from the JPS, the P5-A satellite will depend on its reaction control system (RCS) for orbit maintenance around Mars. It will act as a telecommunication relay satellite for ARCHIMEDES and later for other missions.

10.3.3.1 Trajectory

For the AMSAT P5-A mission, a reference target orbit with a 4 000 km pericentre altitude and a 20 000 km apocentre altitude was chosen which is inclined by 30° against the equator. A plan view of this orbit can be seen in Fig. 10.30. Entering the atmosphere during daylight greatly eases the thermal constraints (see chapter 5.4) and is a requirement for the camera experiment. Therefore, the orbit has to be oriented such that its periapsis and with it the atmospheric entry point lies on the sunlit side of the planet. As a result of that, the apoapsis is eclipsed. A sketch of the de-orbit and deployment strategy is given in Fig. 10.31.

To get the camera heads a scanning view of the environment, the ballute will be left to rotate freely in the air stream and any active means of vortex shedding are left aside. This method necessitates as smooth a skin as possible. This is most important for the equator seam, which might otherwise accidentally act as a small burble fence.

To prolong measurement time in the upper atmosphere, it is planned that the spacecraft will enter the atmosphere of planet Mars multiple times, until it has lost enough speed to remain within the atmosphere for a final descent to the ground. A small risk exists that the vehicle exits the atmosphere on a suborbital trajectory that re-enters half-way around the planet on the night side. This would end the thermal life of the instrument pod after aerodynamic heating cools off and would jeopardize the camera experiment. However, the Monte-Carlo analyses of some 5000 trajectories given in Fig. 10.32 shows that such a risk is acceptable, as most missions complete enough of a full orbit to descend on the sunlit hemisphere.

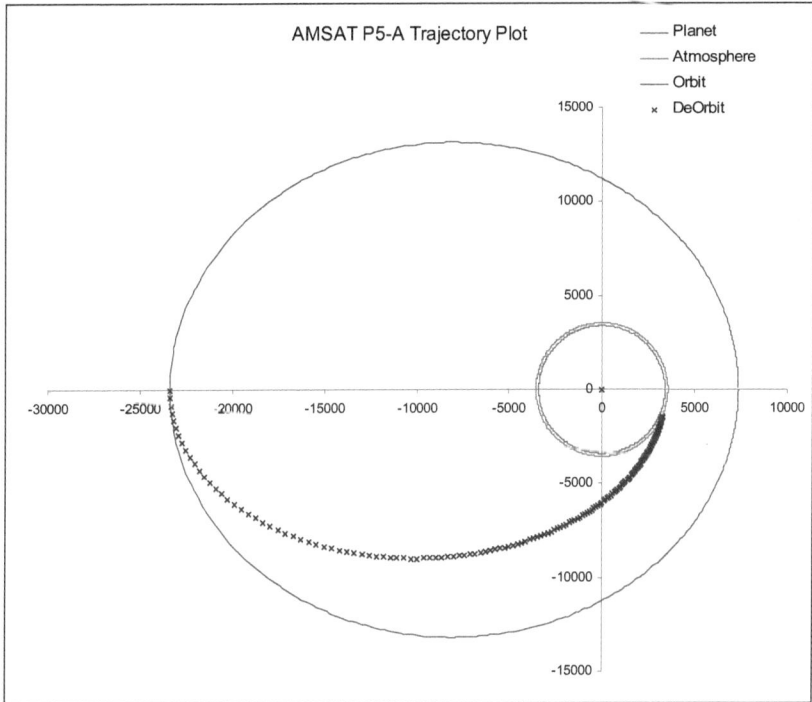

Fig. 10.30: Plot of the intended AMSAT P5-A orbit.

To make ARCHIMEDES as light as possible, the ballute is filled with Helium, which adds significant static lift to aerodynamic drag during final descend. The resulting trajectory allows continuous measurement times of several hours ranging from the outermost atmosphere layers down to the ground.

Since we know from chapter 3.5 (see page 26) that a skipping entry presents a delicate navigation problem, a two-step approach strategy was designed. In this scenario, the JPS performs a total of two or three de-orbit burns. The first one is carried out with the 400N main engine to lower the periapsis of the JPS orbit to around 400km above the atmospheric interface. The orbit is then closely measured through radio ranging from the ground and the orbiter to calibrate the engine running on an almost empty tank system. If required, the manoeuvre can be repeated, to bring the JPS even closer to the atmosphere. The final deorbit impulse can be given either by the precisely calibrated main engine or cold-gas RCS thrusters which have a higher accuracy.

Once the final deorbit impulse has been made, the JPS coasts towards the atmosphere. About three hours out from atmospheric entry, the ballute spacecraft is deployed and inflated and the JPS eventually discarded.

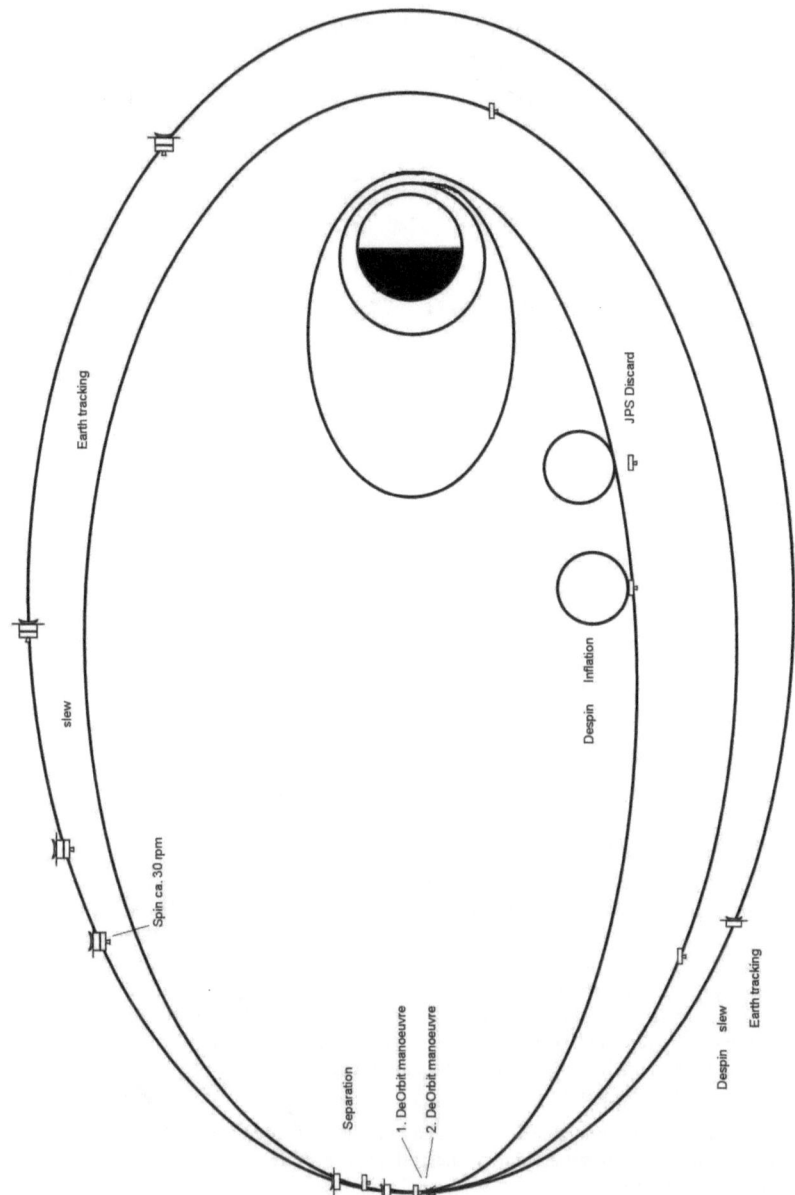

Fig. 10.31: ARCHIMEDES de-orbit and deployment procedure.

The actual mission elapsed time and number of atmospheric entries have been determined by performing an error analysis and by studying "shallow", "nominal" and "steep" cases of a "normal" (34 kg entry mass) and "heavy" (65 kg entry mass) spacecraft. In a "heavy" and "shallow" case, the given spacecraft will enter the atmosphere up to 8 times before final descent, while in a "steep" case the system will decelerate directly onto its decent trajectory, without any further orbit revolution. The nominal case is a 2-entry mission, where the first entry into the atmosphere lowers the apoapsis and the second entry leads to the final descent. Results of these mission analysis, including telecommunication orbiter visibility, are given as examples in chapter 5.2, starting on page 45.

With these results, operating modes can be defined for the three scientifically interesting mission phases (see Table 10.11). While in space, the vehicle operates in SKY-mode, during which pictures and magnetometer readings can be taken. During SKY-mode, the residual outer atmosphere can also be probed by a complement of accelerometers, making use of the large and light configuration, to directly probe very tenuous outer atmosphere layers. SKY-mode is set initially after successful deployment of the spacecraft and terminates when HY mode is set.

During hypersonic (HY) mode, the vehicle will take flow field measurements with the COMPARE instrument, as well as images of the planet horizon (limb sounding) and surface through various optical filters. Because the ballute is tumbling and rolling in the aerodynamic flow, the camera is scanning the surroundings and imaging can be triggered, by determining the current orientation with the help of the accelerometers or a complement of possible optical sensors. HY-mode is set when accelerometers register a predetermined acceleration level (for instance the threshold for atmospheric entry as discussed in 3.5 starting on page 26) and terminates when either the acceleration level falls back below this threshold (minus a hysteresis) or the acceleration signature lets the flight computer to recognize sound barrier transition.

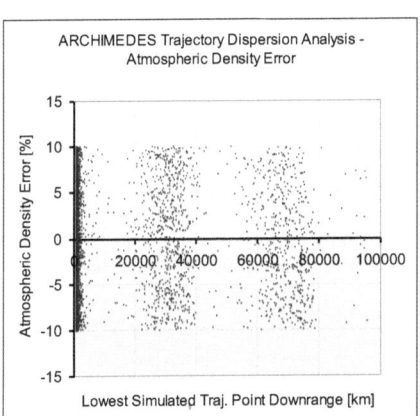 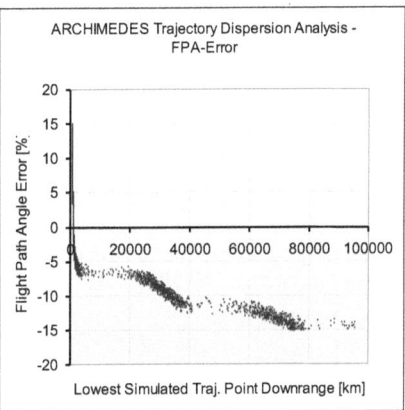

Fig. 10.32: Monte-Carlo analysis, showing the downrange distance of the landing point from the initial entry point for 5000 trajectory simulations.

During the final subsonic mode (LOW mode) the vehicle takes atmosphere readings, the only phase where this is possible, magnetometer readings and images. Again, the rolling spacecraft will make the camera heads scan the surroundings and accelerometers can be used again to determine when to capture an image. LOW mode is set once the NCA is discarded and stays set for the remainder of the battery life.

10.3.3.2 Telecommunication and Power Budgets

Based on these modes, the mission analysis from chapter 5.2 and the scientific sensor suite data from Table 10.5, a data communication plan can now be established, which then completes the power budget.

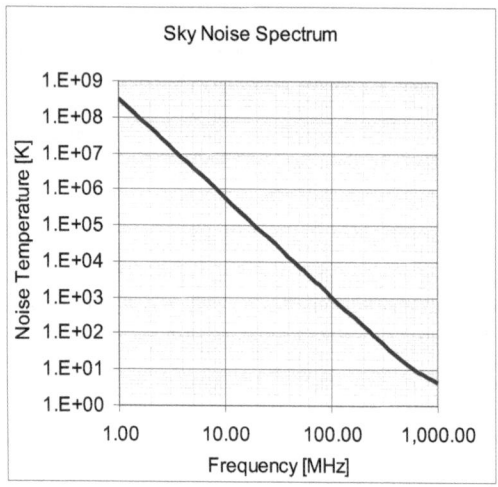

Fig. 10.33: Sky noise spectrum below 1GHz [105].

Since the ARCHIMEDES ballute spacecraft has no means of actively controlling its attitude and flies on a trajectory that is as unpredictable as they come, neither the orbiter nor the ballute spacecraft can know exactly when or in which direction the other becomes visible. This in turn leads to both spacecraft having to rely on omnidirectional antennas for communication.

Data communication theory [106][107] shows that in such a set-up, lower frequencies offer a higher theoretical data throughput. This is compensated by background sky noise, which rises significantly for frequencies below 1GHz [108]. Fig. 10.33 shows the sky noise model below 1GHz as given by the CCIR.

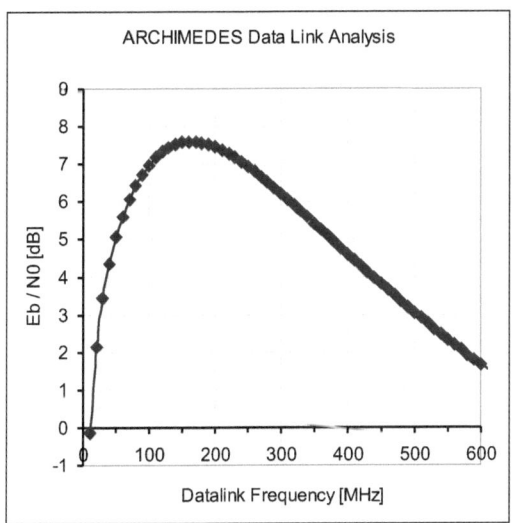

Fig. 10.34: ARCHIMEDES data link analysis assuming BPSK with 90° phase angle and no coding gain on 2.8 kbps over 25 000 km slant range.

These two effects lead to a theoretical data rate maximum around 145MHz. A data communication link in this VHF band offers a roughly 4dB better link budget than the 430MHz UHF band (see Fig. 10.34). The UHF band however is used by all modern interplanetary space probes (such as Mars Express, MRO and MER), employing the Proximity-1 protocol as recommended by CCSDS for inter-spacecraft communications and surface operations [109]. The advantage of the standard is clearly that any Mars orbiting satellite can relay data of any other vehicle on or around Mars, theoretically creating a handsome telecommunication network. So, while offering a better performance in terms of link margin or possible data rate, a VHF communication system would be unique and make it impossible to use any other vehicle as a reliable data communication relay.

In our case of an uncontrolled ballute, it has to be considered that achieving a reasonable data rate across large distances (especially when the telecommunication orbiter is far – 20 000 km apocentre altitude were planned by AMSAT) is already difficult. Reasonable amounts of data also have to be downloaded before and after each hypersonic pass, so that in the event of a spacecraft loss during this critical mission phase, important data is still available. This, in turn, demands timely downloads and does not allow long-time low data rate communication sessions.

Sky-Mode

| Mode Unit Duration | 3,600 s | | 0D 1h 0m 0s | SKY |

	# of Readings	Operation Duration [s]	Total Energy [Ws]	Total Energy [Wh]	Total Compr. Data [Bit]
Camera	2	26.00	26.00	0.007	6,000,000
Magnetometer	3,600	3,600.00	1,800.00	0.500	72,000
Atmos Package	0	0.00	0.00	0.000	0
COMPARE	0	0.00	0.00	0.000	0
AMS	36,000	3,600.00	1,800.00	0.500	792,000
Housekeeping	360	3,600.00			69,840
Flight Computer		3,600.00	2,880.00	0.800	
Power Conditioning		3,600.00	360.00	0.100	
Total			**7,046.00**	**1.957**	**6,933,840**

Hypersonic Mode

| Mode Unit Duration | 300 s | | 0D 0h 5m 0s | HY |

	# of Readings	Operation Duration [s]	Total Energy [Ws]	Total Energy [Wh]	Total Compr. Data [Bit]
Camera	1	13.00	13.00	0.004	3,000,000
Magnetometer	300	300.00	150.00	0.042	6,000
Atmos Package	0	0.00	0.00	0.000	0
COMPARE	15,000	300.00	300.00	0.083	120,000
AMS	3,000	300.00	150.00	0.042	66,000
Housekeeping	300	300.00			58,200
Flight Computer		300.00	240.00	0.067	
Power Conditioning		300.00	30.00	0.008	
Total			**1,783.00**	**0.495**	**3,250,360**

Subsonic Mode

| Mode Unit Duration | 3,600 s | | 0D 1h 0m 0s | LOW |

	# of Readings	Operation Duration [h]	Total Energy [Ws]	Total Energy [Wh]	Total Compr. Data [Bit]
Camera	3	0.01	39.00	0.011	9,000,000
Magnetometer	3,600	1.00	1,800.00	0.500	72,000
Atmos Package	30	0.13	225.00	0.063	16,020
Housekeeping	360	1.00			69,840
Flight Computer		1.00	2,880.00	0.800	
Power Conditioning		1.00	360.00	0.100	
Total			**16,104.00**	**4.473**	**9,159,780**

Table 10.11: ARCHIMEDES flight modes for various mission phases.

To achieve a reasonable data rate in the UHF band, the orbiter could now add a steerable antenna beam (i.e. a mechanical platform or phased array) for some sort of an antenna gain, directing it towards Mars. In order to compensate for the 4dB loss in link margin, this would at least have to be an antenna directing its beam towards one hemisphere. Because the P5-A satellite is foreseen to track Earth with its high gain antenna and solar arrays (see vehicle configuration in Fig. 10.27a), a communication antenna searching for the ballute would have to be able to rotate the beam once around the spacecraft. In case of P5-A, which is a design based on the P3-D amateur radio communication satellite of similar dimensions, changing the primary satellite bus configuration is impractical, due to the existing launch adapter, handling rigs and other hardware.

Data Transmission Geometry		Data Transmission Geometry	
Transmission Range	20,000.00 km	Transmission Range	20,000.00 km
Possible Data Rate	2,800.00 bps	Possible Data Rate	2,800.00 bps

ARCHIMEDES Link Budget VHF		ARCHIMEDES Link Budget UHF		
Transmission Frequency	145.00 MHz	Transmission Frequency	430.00 MHz	
Transmitter Power	3.01 dBW	Transmitter Power	3.01 dBW	2.00 W
Line Loss Tr-2-Ant	-1.00 dB	Line Loss Tr-2-Ant	-1.00 dB	
Transmitter Antenna Gain	0.00 dB	Transmitter Antenna Gain	0.00 dB	
Space Range Loss	-161.70 dB	Space Range Loss	-171.14 dB	
Transmission Path Loss	0.00 dB	Transmission Path Loss	0.00 dB	
Receiver Antenna Gain	0.00 dB	Receiver Antenna Gain	0.00 dB	
System noise Temp	487.36 K	System noise Temp	122.06 K	
EIRP	2.01 dBW	EIRP	2.01 dBW	1.59 W
Eb / N_0 (@ Apocenter)	7.56 dB	Eb / N_0 (@ Apocenter)	4.13 dB	
C / N_0 (@ Apocenter)	42.03 dB	C / N_0 (@ Apocenter)	38.60 dB	

Transceiver		Transceiver	
Efficiency on Transmit	40%	Efficiency on Transmit	40%
Power Input on Transmit	5.00 W	Power Input on Transmit	5.00 W

Table 10.12: ARCHIMEDES to orbiter link budget. Left: Long range VHF to P5-A. Right: Short range UHF to a possible other orbiter. No coding scheme applied.

Additionally, if the ballute spacecraft's communication architecture is not bound to the Proximity-1 protocol, higher order coding schemes can be used, such as the one developed by France Telecom and licensed to AMSAT for exclusive use in their Mars mission [110]. This scheme can get data rate within a few percent of the theoretical Shannon-limit.

For the ARCHIMEDES mission, the telecommunication system was therefore designed for maximum coverage through a radio that can transmit on both frequencies. The high-performance VHF system does not conform to CCSDS standards, using a proprietary amateur radio link for communication with the P5-A orbiter. The standard UHF system uses the Proximity-1 protocol and can therefore communicate with any vehicle in the Martian system that is willing to act as a relay. The UHF system, though offering a somewhat lower overall performance, is usually used with vehicles in a low Mars orbit, therefore compensating through closer range and more frequent visibility (if their operating agencies would agree to use their assets to that effect).

Link budgets for both communication modes are given in Table 10.12.

Overview Power Characteristics	
Main bus voltage	5.00 V
Min. Control drop	1.00 V
Allowed EOL voltage drop	1.00 V
Minimum Bat voltage	**7.00 V**
Maximum Bus Power	10.94 W
Maximum Bus Load	2.19 A
Maximum Discharge Load	1.56 A

Table 10.13: ARCHIMEDES ballute spacecraft overall power supply requirements.

For power and computational reasons (the flight computer has to handle the data in different ways), using both systems simultaneously is currently not planned. The flight computer has to negotiate available data links by itself and choose the most promising one. Simultaneous VHF/UHF communication, however, is still an option to be considered in future studies.

With this information and general power requirements (see Fig. 10.13) derived from the

Nose Cover Assembly Power				Instrument Pod Power			
	Steep	Nominal	Shallow		Steep	Nominal	Shallow
Total Energy [Wh]	45.72	52.53	246.01		72.03	97.62	90.04
Operating Time [h]	6.56	8.65	40.07		12.00	16.32	16.00
Min. Bat Capacity [Ah]	6.53	7.50	35.14		10.29	13.95	12.86
Excess Capacity [Ah]	31.87	30.90	3.26		6.99	3.33	4.42
Available Capacity [%]	5.88	5.12	1.09		1.68	1.24	1.34

Req. Min. Bat. Capacity NCA	35.14 Ah		13.95 Ah

	NCA Battery Pack		Pod Battery Pack
Type	FriWo LiMnO2 M24HR		FriWo LiMn02 M52
Unit Mass	201.00 g		51.00 g
Nominal Unit Voltage	2.80 V		2.80 V
Nominal Unit Capacity	20.00 Ah		4.50 Ah
Self discharge rate (% p. a.)	0.02		0.02
Actual Unit Capacity after 2yrs	19.20 Ah		4.32 Ah
Number of serial packs	3 -		3 -
Number of parallel packs	2 -		4 -
Total Number of Units	6 -		12 -
Pack Voltage	8.40 V		8.40 V
Pack Capacity	38.40 Ah		17.28 Ah
Required Unit Discharge Current	781.67 mA		390.83 mA

Total Battery Mass NCA	1.21 kg		0.61 kg

Table 10.14: ARCHIMEDES Instrument Pod and Nose Cover Assembly power budgets.

scientific sensors and avionics suite, a power budget and power subsystem can be designed for all scenarios and mission phases. This completes the mass budget of the ballute spacecraft.

The power budget summary is shown in Table 10.14.

10.3.3.3 Thermal Environment and Hypersonic Flow

With all masses and sizes in place, trajectories can now be calculated again in greater detail, according to chapter 5.2 and compared against data volume, thermal and flight dynamics requirements and constraints. Results were already given for the configuration herein as examples in chapter 5.2. They show that the mission as such is very well possible.

If a multiple entry scenario is chosen for the sake of extended upper atmosphere measurement time, navigational and thermal design requirements become more difficult to meet. The highly delicate entry can quickly lead to a trajectory with more hypersonic passes than the

	Max g	Max Heating
Flight Mach Number	5.27	22.36
Flow velocity [m/s]	1062.5	4194.9
Deceleration [g]	1.73	0.43
Ambient Pressure [Pa]	100	15
Atm. Temp. [K]	161.65	140.04
convective heating [W/m²]	564	4525

Table 10.15: Sample set of ARCHIMEDES ALPHA1-Norm type entry trajectory points used for CFD analysis.

ballute can handle. The passing of the ballute through Mars' shadow also causes temperatures to drop down to around 150K. This has to be kept in mind when working on the instrument pod's thermal design. This is of particular importance as repeated light-shadow passes cause thermal load cycles not just for the pod but also the ballute and its seams.

A typical trajectory is the Alpha1-norm orbit with two entries. The trajectory points of particular interest are the point of maximum deceleration and the point of maximum heating, key data of which is given in Table 10.15. Corresponding CFD results for flow field temperatures is given in Fig. 10.35.

To investigate the chemical composition of gas in the boundary layer and to determine whether it contains any potentially harmful constituents, the concentration of gasses close to the ballute skin is given in Fig. 10.36. It can be seen that free atomic oxygen has to be dealt with, but also that the concentration is very low, so no significant degradation of the ballute surface is expected for the duration of exposure (several minutes at most).

The pressure on the surface of the ballute was also determined and is given for the point of maximum deceleration (also the point of maximum dynamic pressure) in Fig. 10.37. The pressure across the ballute skin is significantly lower than the nominal inflation pressure, therefore confirming the assumption that the ballute stays well in shape during the entire mission.

This concludes the ARCHIMEDES mission design and feasibility study. A more detailed discussion of the CFD results can be found in reference [42].

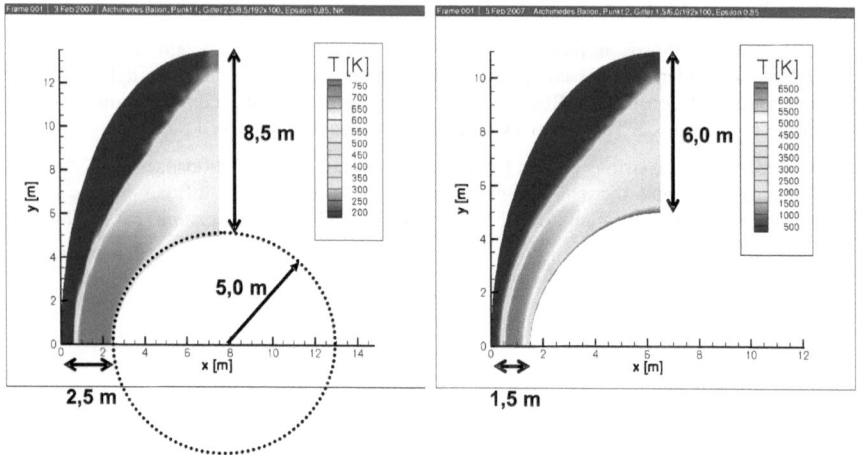

Fig. 10.35: CFD results for max-g (left) and max. convective heating (right) during entry of ARCHIMEDES.

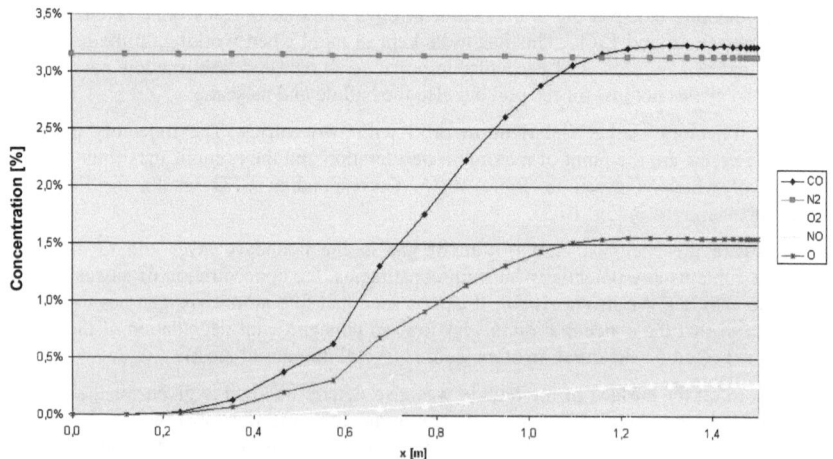

Fig. 10.36: Flow field constituents as obtained from a CFD analysis of the point of maximum heating during atmospheric entry of ARCHIMEDES.

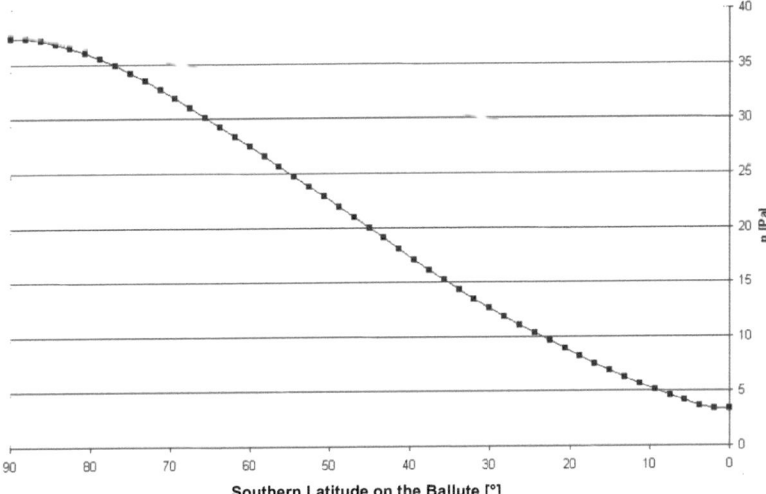

Fig. 10.37: Dynamic pressure distribution across southern latitudes of the ballute for the point of maximum deceleration.

10.3.4 Conclusion

The analysis shows that the ARCHIMEDES mission is possible according to the state of the art and that a scientifically viable mission can be flown. However, some research is necessary regarding the ballute manufacturing process and necessary material procurement (PBO). To the knowledge and conviction of the author, these are no prohibitive problems for a reasonably funded interplanetary probe project. Fig. 10.38 shows the ARCHIMEDES ballute spacecraft as it rolls across the surface of Mars, driven by the winds like a tumble weed. Note the inflation hose extending outward from the north pole of the vehicle.

Fig. 10.38: Artist's impression of the ARCHIMEDES ballute after completion of its mission [96].

10.4 Other Applications

Many space applications would greatly benefit from the inflatable hypersonic drag device known as "ballute". Aside from its use as a scientific probe, it might also be used as an entry vehicle to deliver a payload to the surface of a planet that is inaccessible to conventional parachutes or requires a heat shield too big for available rocket fairing diameter. Planetary aerocapture and aerobreaking applications could also make good use of a ballute, as would an orbital recovery system for astronauts. Tourists into extreme sports, who have done everything in their life except a hypersonic entry from space, might find it worth the expense to fly a ballute from a suborbital trajectory to the ground. Eventually, it is even conceivable to build a solar sail ballute. For good measure his chapter will give very brief introductions into a scientific probe for Venus concept, as well as an orbital recovery system.

10.4.1 The ARCHIMDES-V Venus Ballute Probe

The proposed mission design of ARCHIMEDES-V (Venus) [111] is similar to that of the AMSAT P5A-ARCHIMEDES design (see chapter 10.3) – it foresees the use of a Venus orbiter as an interplanetary carrier spacecraft that enters into orbit around the planet prior to probe release. The carrier spacecraft will also have to serve as a telecommunication bridge with planet Earth and relay probe data between the planets. To be able to measure atmospheric data already in high altitude, the system will enter the atmosphere fully inflated and, if so desired, enter even multiple times before finally descending all the way to the ground. In contrast to the Mars mission, the balloon will be designed such that it will be superpressurized above an altitude of about 50km and collapse thereafter under the atmospheric pressure increase. Using this approach, the balloon will have a relatively constant descent speed that can be tailored to the mission requirements, by adjusting the initial gas mass and the balloon diameter.

The scientific instruments of ARCHIMEDES-V shall reside inside a carbon-fibre reinforced titan pressure vessel, together with primary batteries for power supply. The heat dissipated from the instruments' electronic systems is absorbed by the thermal subsystem consisting of a phase-change fluid system that changes from solid to liquid, or just of a large aluminum block, depending on mission design. The pressure vessel shall be isolated against the Venusian heat by glass fibre insulation blankets and an exterior Meta-Aramid fibre sheet. To minimize heat transfer from the exterior of the vessel to the instruments, they are suspended by elements with low thermal conductivity and isolated against the pressure vessel's interior walls by vacuum and appropriate coated shells. Protection from aerothermal heating is provided by a nosecone assembly with a CC/SiC heat shield that can house additional instruments, as well as additional batteries. The nosecone assembly will be jettisoned after the flight Mach number drops below 0.8, but dropping it at higher Mach numbers (up to 2.5) may be considered, if scientifically desirable.

10.4.2 Sky Raft: Recovery from Earth Orbit

Ballutes may also be used to recover objects from Earth orbit that would otherwise be destroyed during entry. Such an object could be a payload that was initially not intended to return to the surface of Earth and cannot be brought aboard existing re-entry vehicles. On the dramatic side, such an object could also be an astronaut (or a crew of astronauts) who ended up in a perilous situation.

A number of predicaments can occur in space. For example, Gemini astronaut Ed White almost didn't make it back when the door of his spacecraft couldn't be locked shut behind him as planned [112] In another scenario, a crew of astronauts sit inside a spaceship which is unable to safely return to the surface of the Earth, such as it was the case with Space Shuttle Columbia.

The necessity to deal with such a situation and the advantages of a ballute system as means of rescue were already discussed by engineers at Caltech's JPL [113]. In this particular case, a system to rescue 7 astronauts in a single cabin was considered, as a result of the loss of Shuttle Columbia.

The solution discussed herein addresses an ejection seat type emergency rescue ballute that functions like a life raft in the sky. The concept was therefore named Sky Raft.

It is a self sufficient subsystem intended to rescue a single astronaut or a crew of astronauts from life threatening situations in low Earth orbit (LEO) and return them safely to the surface. The package contains a ballute, a service system and a rescue pod and therefore resembles the general ballute systems described herein. The ballute is a spherical 5.7 metre diameter balloon made of a NOMEX-Zylon fabric composite with a gas tight inner liner. To minimize roll and tumble movements, a burble fence or rows of stripes should cause a uniform shedding of vortexes. The rescue pod can either be an enclosed cabin in which the astronauts lie or sit, or an attachment fixture for an external space- or flight suit. The service system contains the gas generator-augmented IGSS with a Helium main charge fed from a pressure bottle, a complement of attitude sensors, cold gas thrusters, possibly a de-orbit motor and a flight computer.

The astronaut would carry the Sky Raft package as part of some type of manned manoeuvring unit, his space suit back pack or integrated into an ejection seat. If predicament is positively established and the astronaut wishes to abort his mission to the surface, he activates the Sky Raft (and ejection seat). The raft control computer then establishes the astronaut's attitude and orbit and manoeuvres him onto an approach trajectory by firing a small solid rocket motor and a set of simple cold gas thrusters for trajectory fine-tuning and attitude control. Depending on the type of emergency, the return can happen immediately or target a favourable landing area. If timely return to Earth is of the essence, like when life support resources are running low or the suit is loosing coolant, anywhere on the surface of planet Earth might offer higher chances of survival than remaining in space. Otherwise weather, hostile countries or remote places can be avoided.

Once the entry trajectory has been established, the ballute is deployed and inflated. The service system, ejection seat and other unused systems are discarded, while the astronaut is left to enter the atmosphere tied to the rescue pod. The spacesuit is clamped to the rescue pod at the belly section, so the astronaut enters the atmosphere with his back forward. The human body can withstand the highest g-loads when lying on his back. The suit itself can easily be made to withstand entry heating in the way a fire fighter's uniform is made and as with IRDT, several layers might even be allowed to burn off during entry. The astronaut must also slide his feet, arms and his head into one or several pockets underneath the ballute's south pole to give them additional support during high-g deceleration and to make the astronaut more aerodynamically smooth. An additional padded seat back-plate made from contoured CC/SiC might be left attached to the atsronaut, if heating proves difficult to handle. Unfortunately, in such a simple configuration, the heavy astronaut will always roll forward into the hypersonic flow, no other configuration is stable.

Close to the ground, the Sky Raft ballute can be released. Because of its Helium filling, it will quickly separate from the falling astronaut, who can now deploy a conventional parachute for final, controlled descent.

A Sky Raft of this design would subject the astronaut to a deceleration of up to 8 g and a stagnation point surface temperature of up to 1000 °C assuming an entry with an angle of -1.4° from a 400 km circular orbit. It would add around 130 kg to the astronaut's equipment.

11 Conclusion and Outlook

The study shows that existing balloon technology, as well as existing inflation systems technology, can be combined to create a robust inflatable drag body (referred to as "ballute") that is able to withstand the mechanical and heating loads during atmospheric entry. It could also be shown that a high performance ballute may be used to achieve a certain scientific goal by reaching altitudes otherwise inaccessible (which is a new concept introduced with this thesis). It also shows that such a ballute spacecraft is technically possible.

An underlying theory and a design guide for such a vehicle was developed. It can be used for the study of a ballute spacecraft in general, regardless of purpose. It was used herein for the design of the suggested Mars mission ARCHIMEDES and its flight test vehicle MIRIAM.

The theory shows that with available film gauges, a viable scientific mission can be flown with sufficient ballute payload mass for a meaningful scientific instrument package (see Fig. 11.1).

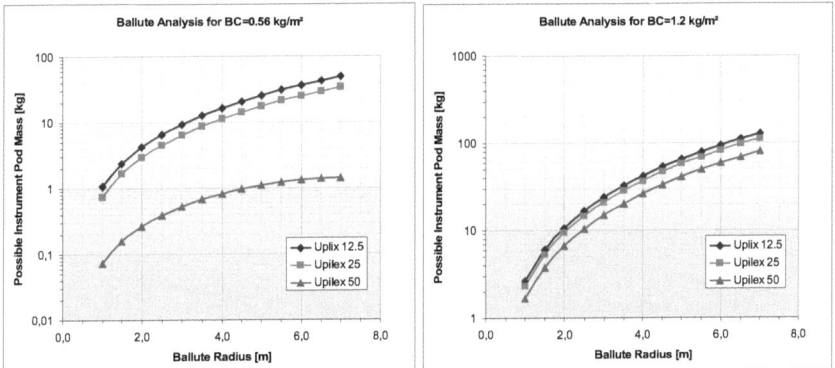

Fig. 11.1: Possible Instrument Pod mass as a function of ballute radius for various UPILEX films and desired ballistic coefficients (without considering material strength).

Material analysis showed that thin films with these gauges exist that are sufficiently strong, but that procurement might be an issue with the most suitable candidate material (PBO thin film). It was also shown that manufacturing and seaming methods need further development. Adhesive tapes alone are insufficient and welding techniques, which deliver very good results, are fairly new and unexplored. Appropriate manufacturing equipment for welded seams is bulky, complex and has to be custom made according to testing results.

How well the theory works in practice could not be verified. The flight test system MIRIAM never got deployed properly after a bolt malfunctioned in space. As this test is a necessary confirmation of the technical concept put forward herein, it is strongly recommended to repeat it, before any further studies of the matter take place.

It could already be shown, however, that a ballute for space applications can be developed and that it is a very real as well as promising alternative technology option for atmospheric entry.

Bibliography

[1]   Grieger,B., et al., "The ARCHIMEDES Mission Science Proposal", Mars Society Deutschland e.v. & Max-Planck Institut für Sonnensystemforschung, 2005

[2]   F.Forget, F.Costard, P.Lognonné, "Planet Mars, Story of Another World", Springer, 2006

[3]   M. Pätzold, S. Tellmann, B. Häusler, D. Hinson, R. Schaa, G. L. Tyler, "Discovery of a Sporadic and Local Third Layer in the Ionosphere of Mars", Science, Vol. 310, Issue 5749, pp. 837-839, 2005

[4]   D. P. Hinson, M. Pätzold, R. J. Wilson, B. Häusler, S. Tellmann, and G. L. Tylor, "Radio occultation measurement and MGCM simulations of Kelvin waves on Mars", Icarus 193, pp. 125-138, 2008

[5]   H.S.Griebel, "Methoden der Ballonentfaltung auf dem Mars: Konzepte zur praktischen Durchführung und Versuchsanordnungen zu deren Erprobung RT-DA 03/01", Technical University of Munich, 2002

[6]   D.H.Lorenzen, "Mission: Mars - Die sensationellen Entdeckungen der neuen Raumsonden", Kosmos, 2004

[7]   Jacques Blamont, "Planetary Balloons", Springer, 2008

[8]   "ARES - A Proposed Mars Scout Mission", http://marsairplane.larc.nasa.gov/, 2009

[9]   V.V.Kherzanovich, J.Cutts et.al., "The Mars Aerobot Validation Program", JPL, 1999

[10] Mark Grahne, David Cadogan, "Deployment Control Mechanisms", Gossamer Spacecraft: Membranes and Inflatable Structures Technology for Space Applications, AIAA, 2001

[11] J.P.Talentino, "Development of the Fabrication and Packaging Techniques for the ECHO II Satellite", NASA, 1966

[12] D.L.Clemmons, Jr., "The Echo I Inflation System, NASA TH D2194", NASA, 1964

[13] D.Wilde, S.Walther, K.Pitchadze wt. al., "Inflatable Reentry and Descent Technology (IRDT) – Further Developments", 2nd International Symposium of Atmospheric Reentry Vehicles and Systems, Arcachon, France, 2001

[14] D.Wilde, S.Walther, K.Pitchadze wt. al., "Flight Tests and ISS Application of the Inflatable Reentry and Descent Technology (IRDT)", Acta Atsronautica Vol 51, No.1-9, 2002

[15] General Motors, "Buick, Cadillac, Oldsmobile Air Cushion Restraint System Service Manual", General Motors Corporation, 1974

[16] Spartan Project Team, "Spartan 207/Inflatable Antenna Experiment Preliminary Mission Report", NASA Goddard Space Flight Center, 1997

[17] "The Convenience and Security of Cadillac's New Air Cushion Restraint System", General Motors Corporation, 1974

[18] Brian O'Neill, "There Is Life in Those Old Cars Yet", New York Times, February 7, 1993

[19] M. de Jong, A. Lennon, "Pressure Restraint Design for Inflatable Space Habitats", Thin Red Line Aerospace Corp., 2007

[20] James R. Hansen, "SP-4308 Spaceflight Revolution", NASA Langley Research Center, 1994

[21]   Raymond Turk, "Pressure Measurements on Rigid Model of Ballute Decelerator at Mach Numbers from 0.56 to 1.96", NASA, 1966

[22]   Theo Knacke, "Parachute Recovery Systems Design Manual", Para Publishing, 1992

[23]   James M. Grimwood, Barton C. Hacker, Peter J. Vorzimmer, "Project Gemini Technology and Operations", NASA, 1968

[24]   Angus McRonald, "A light Weight Inflatable Hypersonic Drag Device for Planetary Entry", JPL, 1999

[25]   Jeff Hall, "A Review of Ballute Technology for Planetary Aerocapture", NASA, JPL, 2000

[26]   Peter Gnoffo, Brian Anderson, "Computational Analysis of Towed Ballute Interactions", AIAA, 2002

[27]   Frank J. Regan, Satya M. Anandakrishnan, "Dynamics of Atmospheric Re-Entry", AIAA, 1993

[28]   John Anderson, "Hypersonic and High-Temperature Gas Dynamics 2nd Ed.", AIAA, 2006

[29]   K.Gersten, "Einführung in die Strömungsmechanik", Vieweg, 1991

[30]   P.H.Zipfel, "Modeling ans Simulation of Aerospace Vehicle Dynamics", AIAA, 2000

[31]   ANSI, AIAA, "Guide to Reference and Standard Atmosphere Models G-003B-2004", ANSI, AIAA, 2004

[32]   Forget, Hourdin, Fournier, Hourdin, Talagrand, Collins, Lewis, Read, Huot, "Improved general circulation models of the Martian atmosphere from the surface to above 80 km", J.Geophys. Res., 104(E10), 24,155–24,175, 1999

[33]   G.W.Prölss, "Barometrische Höhenformel", Physik des Erdnahen Weltraums, pp.37-40, Springer, 2004

[34]   David R. Williams, "Planetary Fact Sheet Mars", NASA Goddard Space Flight Center, 2007

[35]   N.X.Vinh, J.R.Johanessen, J.M.Longuski, J.M.Hanson, "Second-Order Analytic Solutions for Aerocapture and Ballistic Fly-Through Trajectories", Journal of the Astronautical Sciences, 1984

[36]   Kristin Gates Medlock, James M. Longuski, "An Approach to Sizing a Dual-Use Ballute for Aerocapture, Descent, and Landing", Purdue University, 2006

[37]   K.Sutton, R.A.Graves jr., "A General Stagnation-Point Convective-Heating Equation for Arbitrary Gas Mixtures", Technical Report: NASA TR R-376, 1971

[38]   R.A.Micheltree, M. DiFulvio, T.J. Horvath, R.D.Braun, "Aerothermal Heating Predictions for Mars Microprobe", Technical Report: NASA-AIAA-98-0170, 1998

[39]   M.Zuban, "Untersuchung des Atmosphäreneinflusses auf Raumflugbahnen im Radio Science Simulator des Instituts für Raumfahrttechnik", Institut für Raumfahrttechnik, Universität der Bundeswehr München, 2005

[40]   C. Fiebig, "Erstellen eines ersten Thermalmodells für das Mars BallonExperiment Archimedes im Weltraum", Institut für Raumfahrttechnik, Universität der Bundeswehr München, 2005

[41]   L.Kotte, "Thermalanalyse der Mars Erkundungssonde ARCHIMEDES", Institut für Raumfahrttechnik, Universität der Bundeswehr München, 2007

[42]   T.Schramm, "Untersuchung der aerothermodynamischen undmechanischen Wechselwirkung amMars-Ballon ARCHIMEDES", Institut für Thermodynamik, Universität der Bundeswehr München, 2007

[43]   H.Pasemann, "Implementierung von Slip-Randbedingungen in das Programm CEVCATS", Institut für Thermodynamik, Universität der Bundeswehr München, 2004

[44] M. Zuban, "Untersuchung des Einflusses der Hyperschallströmung auf die Form eines Ballons beim Eintritt in dieMarsatmosphäre", Institut für Raumfahrttechnik, Universität der Bundeswehr München, 2006

[45] T.Högner, "Aeroelastische Analyse des ARCHIMEDES Ballons", Institut für Raumfahrttechnik, Universität der Bundeswehr München, 2008

[46] J.Selle, "Planung und Simulation von Radio-Science-Experimenten Interplanetarer Raumfahrt-Missionen", Institut für Raumfahrttechnik, Universität der Bundeswehr München, 2005

[47] B.Häusler, O.Zeiler, G.Billig, W.Eidel, "Satellitensimulator für Kleinsatelliten", Institut für Raumfahrttechnik, Universität der Bundeswehr München, 1996

[48] Jaremenko, I.M., "BALLUTE Characteristics in the 0.1 to 10 Mach Number Speed Regime", Journal of Spacecraft and Rockets, 1967

[49] T.Krauledat, "Analyse und Simulation des Missionsablaufs der Raumsonde Archimedes", Institut für Raumfahrttechnik, Universität der Bundeswehr München, 2006

[50] Mundt,Ch., Griebel,H., Welch,Ch., "Studies of the atmospheric entry of vehicles with very low ballistic coefficient", 13th AIAA/CIRA International Space Planes and Hypersonic Systems and Technologies Conference, 2005

[51] W.G.Vincenti, C.H.Kruger, Jr., "Introduction to Physical Gas Dynamics", John Wiley & Sons, 1965

[52] C.R.Nave, "HyperPhysics, CD ROM Version", Georgia State University, 2006

[53] Harald Hoffmann, Personal conversation and e-mail correspondence on unpublished work, 2005

[54] W.Beitz, K.-H.Küttner, "Dubbel Taschenbuch für den Maschinenbau, 18.Auflage", Springer, 1995

[55] European Cooperation for Space Standardization, "Data for selection of space materials and processes", ECSS Secretariat, ESA-ESTEC, 2004

[56] European Cooperation for Space Standardization, "Thermal vacuum outgassing test for the screening of space materials", ECSS Secretariat, ESA-ESTEC, 2000

[57] K.Bayler, "Archimedes Planetary Protection Specification, V002", The Mars Society Germany (Mars Society Deutschland e.V.), 2003

[58] A. Kunze, F. Epperlein, B. Kröplin, "MARS BALLOON ENVELOPE STUDY", ESA Contract No. 16212/02/NL/PA, 2002

[59] Andre Yavrouian, "High Temperature Materials for Venus Balloon Envelopes", Clearwater, Florida, 1995

[60] A. Yavrouian,; G. Plett,; S. Yen,; J. Cutts,; D. Baek, "Evaluation of Materials for Venus Aerobot Applications", AIAA Summer Conferences, June 28-July 1, 1999

[61] L.Rubin, J.Larouco, "High Performance Balloon Envelope Materials for Planetary Aerobots", Foster-Miller, 2002

[62] "UPILEX-RN Sales Brochure and Technical Datasheet", UBE Europe, 2005

[63] C.Fiebig, "Untersuchung zur Materialschädigung von hochfesten Polyimiden im Hinblick aufTemperatur- und zyklische Beanspruchung", Institut für Thermodynamik, Universität der Bundeswehr München, 2007

[64] Niklas Mesterheide, "Ermittlung der Festigkeitseigenschaften der für die ARCHIMEDES-Mission vorgesehenen UPILEX-Folie unter dynamischer Belastung", Institut für Leichtbau, Universität der Bundeswehr München, 2008

[65] J.Williamson, S.Hetzel, "Effects of high Temperature and Vacuum on Upilex-S Mechanical Properties, TOS-QMC report 2003 / 135", Materials Physics and Chemistry Section, Materials and Processes Devision, ESA ESTEC, 2003

[66]   J.R.Williamson, "Effect of high Temperature on the Tensile Properties of UPILEX and Kapton, TOS-QMC report 2004/026", Materials Physics and Chemistry Section, Materials and Processes Devision, ESA ESTEC, 2004

[67]   G. van Papendrecht, "Results Micro-VCM test E532, TOS-QMC Report 2003-143", Materials Physics and Chemistry Section, Materials and Processes Devision, ESA ESTEC, 2002

[68]   Technical Data Sheet, "UPILEX-S Thermal Control Films", UBE Industries, 2002

[69]   K. Fukuzawa, A. Ohnishi, and Y. Nagasaka, "The Total Hemispherical Emittance of Polyimide Films for Space Use in the Temperature Range 173 K to 700 K", Fourteenth Symposium on Thermophysical Properties,June 25-30, 2000, Boulder, Colorado, USA., 2000

[70]   "PBO Fiber Zylon Technical Information", Toyobo Co., Ltd., Japan, 2004

[71]   E.Orndoff, "Development and Evaluation ofPolybenzoxazole Fibrous Structures, Technical Memorandum 104814", NASA Johnson Space Center, 1995

[72]   U.S. District Court of Columbia, "United States of America vs. Second Chance Body Armor, CV No. 04-0280", 2005

[73]   Toyobo Co. Ltd., "Toyobo Statement on Zylon® as a Ballistic Fiber Used by Body Armor Manufacturers", 2005

[74]   E.Wolf, "Untersuchung über die hochfeste Polyimidfolie aus Polybenzoxazol", Institut für Werkstoffkunde, Universität der Bundeswehr München, 2008

[75]   Anke Veinceur, "Untersuchung zum Alterungsverhalten von Polybenzoxazol mittels Dynamisch Mechanischer Analyse und Fourier-Transformierter Infrarot Spektralanalyse", Institut für Werkstoffkunde, Universität der Bundeswehr München, 2008

[76]   C.Gammel, "Schädigungsuntersuchung an PBO", Institut für Werkstoffkunde, Universität der Bundeswehr München, 2009

[77]   Chul Park, "Theory of Idealized Two-Dimensional Ballute in Newtonian Hypersonic Flow", NASA Ames Research Center, 1988

[78]   F.Mayinger, "Adiabate Gas-Flüssigkeits-Strömungen", Strömung und Wärmeübergang in Gas-Flüssigkeitsgemischen, p.38, Springer, 1982

[79]   Klaus Hornung, Personal Conversation, 2006

[80]   D.Richeson, "Euler's Gem - The Polyhedron Formula and the Birth of Topology", Princeton University Press, Oxford, 2008

[81]   H.Kenner, "Geodesic Math and How To Use It", University of California Press, 1976

[82]   B. van Loon, "Geodesic Domes", Tarquin Publications, 1994

[83]   Helmut Rapp, "Leichtbau 2" Lecture, 2008

[84]   Johannes Wiedemann, "Leichtbau Band 2: Konstruktion", Srpinger, 1989

[85]   "Norton PFA Fluoropolymer ", Saint Gobain Plastics, 2008

[86]   Barbara Bierbaum, "Untersuchungen zur Festigkeit und Lebensdauer von fügetechnisch hergestellten Polyimidverbunden", Institut für Werkstoffkunde, Universität der Bundeswehr München, 2008

[87]   "Norton FluoroWrap FH FEP / Polyimide Composites", Saint Gobain Plastics, 2008

[88]   "Norton FEP Fluoropolymer Film", Saint Gobain Plastics, 2008

[89]   "Norton TH Series Polyimide Film", Saint Gobain Plastics, 2008

[90]   Christoph Wirtz, Saint Gobain Plastics, e-mail correspondence, 2008

[91]   Manuel Windsperger, "Untersuchungen über den Einbau von Instrumentierung in den Folienbalon des Raumflugversuchs MIRIAM", Universität der Bundeswehr München, Universität der Bundeswehr München, 2007

[92] Michael Holz, "Untersuchungen von Naht- undFügetechniken für den Folienballon des Raumflugversuchs MIRIAM", Institut für Raumfahrttechnik, Universität der Bundeswehr München, 2007

[93] Sebastian von Rekowsky, "Konfigurationsuntersuchungen zum Raumflugversuch MIRIAM", Institut für Raumfahrttechnik, Universität der Bundeswehr München, 2007

[94] T.Matzies, H.Rapp, H.Griebel, "Structural Design and Test of the Atmospheric Reentry-Probe MIRIAM", ECIT 2008, Iasi, Romania, 2008

[95] Michael Schwarze, "Erstellen einer 3D-Animation für die geplante Marsmission ARCHIMEDES", Institut für Raumfahrttechnik, Universität der Bundeswehr München, 2009

[96] O.C.Petrache, H.S.Griebel, "Evaluation of the ARCHIMEDES Balloon Deployment in Zero Gravity during the Parabolic Flight Test Campaign VP49", Systems for Space Applications, pp. 23-37, Performantica, Iasi, 2006

[97] H.S.Griebel, "ARM Test System for Parabolic Flight VP 49Technical Manual", Mars Society Deutschalnd e.V., Institut für Raumfahrttechnik, Universität der Bundeswehr München, 2005

[98] H.S.Griebel, "ARM Test System for Parabolic Flight VP 49 Flight User Manual", Mars Society Deutschalnd e.V., Institut für Raumfahrttechnik, Universität der Bundeswehr München, 2005

[99] C.Fiebig, "Optimierung des Kamerafeldes für REGINA", Institut für Photogrammetrie und Kartographie, Universität der Bundeswehr München, 2006

[100] Anna Barth, "Detailierte Auslegung des Ballon-Aufblassystems für den Raumflugversuch MIRIAM.", Institut für Raumfahrttechnik, Universität der Bundeswehr München, 2007

[101] Anna Barth, "Durchführung von Aufblassystemtests für den Raumflugversuch MIRIAM", Institut für Raumfahrttechnik, Universität der Bundeswehr München, 2008

[102] S.Camenzind, "Integration des Erdatmosphärenmodells GRAM-99 in den Radio-Science-Simulator (RSS)", Institut für Raumfahrttechnik, Universität der Bundeswehr München, 2007

[103] L.Roscher, "Aerothermodynamische Analyse des Wiedereintritts für das Höhenforschungsraketenexperiment MIRIAM", Institut für Thermodynamik, Universität der Bundeswehr München, 2008

[104] M.Zähringer, H.Griebel, "A Low Power Radio Altimeter For the ARCHIMEDES Balloon Mission to Mars", ECIT2008, Iasi, Romania, 2008

[105] L.J.Ippolito, "Radio Noise in Satellite Communications", Radio Wave Propagation in Satellite Communications, pp. 123-138, 1986

[106] CCIR, ITU, "Calculation of Free Space Attenuation", Propagation in Non-Ionized Media, pp. 3-8, 1982

[107] CCIR, ITU, "The Concept of Transmission Loss for Radio Links", Propagation in Non-Ionized Media, pp. 8-11, 1982

[108] CCIR, ITU, "Propagation Data Required for Space Telecommunication Systems", Propagation in Non-Ionized Media, pp. 331-350, 1982

[109] Consultative Committee for Space Data Systems, "PROXIMITY-1 SPACE LINK PROTOCOL CCSDS 210.0-G-1", CCSDS Secretariat, Office of Space Communication (Code M-3), Washington DC, 2007

[110] K.Meinzer, Personal Correspondence, 2004

[111] Mundt,Ch., Griebel,H., "Venus Balloon mission Archimedes-V with emphasis on surface and gas radiation phenomena", 1st International ARA days − Atmospheric Reentry Systems, Missions & Vehicles, 2006

[112]   R.Machell, "Summary of Gemini Extravehicular Activity, NASA SP-149", NASA Houston, TX, 1967

[113]   J.Jones, J.Hall, J.J.Wu, "Inflatable Emergency Atmospheric-Entry Vehicles", NASA, JPL, 2004